Finite-Time Stability

Wiley Series in Dynamics and Control of Electromechanical Systems

Finite-Time Stability

An Input-Output Approach

Francesco Amato
University of Catanzaro Magna Græcia
Italy

Gianmaria De Tommasi
University of Naples Federico II
Italy

Alfredo Pironti
University of Naples Federico II
Italy

The right of Francesco Amato, Gianmaria De Tommasi, and Alfredo Pironti to be identified as the authors of this work has been asserted in accordance with law.

Registered Offices
John Wiley & Sons, Inc., 111 River Street, Hoboken, NJ 07030, USA
John Wiley & Sons Ltd, The Atrium, Southern Gate, Chichester, West Sussex, PO19 8SQ, UK

Editorial Office
The Atrium, Southern Gate, Chichester, West Sussex, PO19 8SQ, UK

For details of our global editorial offices, customer services, and more information about Wiley products visit us at www.wiley.com.

Wiley also publishes its books in a variety of electronic formats and by print-on-demand. Some content that appears in standard print versions of this book may not be available in other formats.

Library of Congress Cataloging-in-Publication Data:

Names: Amato, Francesco, author. | De Tommasi, Gianmaria, author. | Pironti, Alfredo, author.
Title: Finite-time stability : an input-output approach / Francesco Amato, University of Catanzaro Magna
    Græcia, IT, Gianmaria De Tommasi University of Naples Federico II, IT, Alfredo Pironti,
    University of Naples Federico II, Italy.
Description: First edition. | Hoboken, NJ : John Wiley & Sons, Inc., 2018. | Series: Wiley series in dynamics
    and control of electromechanical systems | Includes bibliographical references and index. |
Identifiers: LCCN 2018016200 (print) | LCCN 2018017626 (ebook) | ISBN 9781119140566 (pdf) |
    ISBN 9781119140559 (epub) | ISBN 9781119140528 (cloth)
Subjects: LCSH: Stability. | System design.
Classification: LCC QA871 (ebook) | LCC QA871 .A43 2018 (print) | DDC 515/.392–dc23
LC record available at https://lccn.loc.gov/2018016200

Cover design: Wiley
Cover image: © agsandrew/GettyImages

Set in 10/12pt WarnockPro by SPi Global, Chennai, India

Printed and bound by CPI Group (UK) Ltd, Croydon, CR0 4YY

10 9 8 7 6 5 4 3 2 1

To my mother
F. A.

To my family, for all the time I've subtracted to their love
G. D. T.

To Teresa and Andrea
A. P.

Contents

Preface

The concept of finite-time stability (FTS) is useful to study the behavior of dynamical systems within a finite-time horizon. This concept permits to specify bounds on the state and/or the output of a dynamical system, given a bound on its initial state, and/or to constrain the input to belong to a specific class of signals. It follows that finite-time stability is an attractive concept from the engineering point of view, since it gives the possibility to quantitatively specify the transient response of a dynamical system to exogenous inputs (disturbances).

FTS was first introduced in the Russian literature more than sixty years ago [1–3]. The original definition dealt with the *state response of autonomous systems*: a system is said to be *finite-time stable* if, given a bound on the initial condition, its state does not exceed a certain threshold during a specified time interval. During the sixties and seventies, FTS appeared also in the Western literature [4–6], together with the related concept of *practical stability*. This pioneering works, although developing a nice theoretical framework, did not provide computationally tractable conditions for checking the FTS of a given dynamical system, unless simple cases were considered. Therefore, for a long period, this field of research was neglected by control scientists.

At the end of the last century, the development of the Linear Matrix Inequality theory (LMI, [7]) has fueled new interest in the field of finite-time control. In particular, starting from the beginning of the twenty-first century, FTS and finite-time stabilization have been investigated in the context of linear systems (e.g., [8–14]). According to this modern approach to FTS, conditions for analysis and design are provided in terms of feasibility problems involving both LMIs [7] and Differential Linear Matrix Inequalities (DLMIs, [15]), or in terms of solutions of Differential Lyapunov Equations (DLEs, [16]).

As far as *state* FTS is concerned, an effort has been made in order to extend the results obtained for linear systems to the context of nonlinear systems (e.g., [12, 17, 18]), hybrid systems ([19–23]), and stochastic systems ([18, 23–29] among others).

In order to extend the finite-time stability concept to the input-output case, the definition of *input-output finite-time stability (IO-FTS)* was originally given by the authors in [30, 31]. A dynamical system is said to be *input-output finite-time stable* if, given a class of input signals *bounded* over a specified time horizon, the output of the system does not exceed an assigned threshold during the considered time interval. IO-FTS extends the finite-time stability framework to the case of non-autonomous dynamical systems, giving the possibility to set quantitative constraints on the transient response to disturbances.

Just as *state* FTS is an independent concept with respect to Lyapunov stability, also IO-FTS is not related to *classic* IO-stability [32]. The main differences between classic IO-stability and IO-FTS are that the latter involves signals defined over a finite-time interval, does not necessarily require the input and output to belong to the same class, and that quantitative bounds on both input and output must be specified.

The material presented in this book collects and extends the results published by the authors since 2010 on the major control system journals. Besides presenting the main theoretical results to solve both the IO-FTS analysis and synthesis problems for different classes of dynamical systems, a number of case studies are presented as examples of practical applications of finite-time control techniques. Numerical issues related to the solution of DLMIs feasibility problems that arise in the proposed finite-time theory are also discussed, in order to give some guidance to their practical solution.

Chapter 1 introduces the considered finite-time stability framework and presents some preliminary background results that are exploited throughout this monograph. Necessary and sufficient conditions to check IO-FTS for linear systems are provided in Chapter 2, while Chapter 3 deals with the solution of the stabilization (i.e., synthesis) problem. IO-FTS of linear system with nonzero initial conditions is considered in Chapter 4, while the case of IO-FTS with additional constraints on the control input is discussed in Chapter 5. Robust and mixed finite-time/$\mathcal{H}_\infty$ control is presented in Chapter 6, which concludes the discussion concerning the case of linear dynamical systems. The extension of the IO-FTS concepts to a special class of hybrid systems, namely the impulsive dynamical linear systems, is addressed in Chapters 7 and 8; the case of uncertain resetting times for this type of discontinuous dynamical systems is also considered in Chapter 9.

It is important to remark that the IO-FTS approach is useful to refine the system behavior during the transient phase, while classical IO (Lyapunov) stability is a fundamental requirement to guarantee the correct behavior at steady state; therefore, it is a good practice to satisfy both requirements when designing a control system. To this end, in Chapter 10, we illustrate a hybrid architecture, where the controller is implemented by both finite-time control techniques and the classical robust control approach.

The book is completed by five appendices. Appendices A and B provide some preliminary results on LTV systems and matrix algebra; Appendix C illustrates some numerical techniques to solve optimization problems with D/DLMIs constraints, while some MATLAB® scripts that solve this type of optimization problems are presented in Appendix D. Appendix E discusses some real-world examples where the IO-FTS approach can be exploited.

There are some issues that are not presented in this book, in particular those ones that are currently in progress. For example, we do not discuss the extension of the IO-FTS theory to nonlinear, as well as stochastic systems and systems with delays. Here, impulsive systems are only considered from the deterministic point of view, while there is a growing interest for impulsive and switched systems regulated by stochastic phenomena; for such topics the interested reader is referred to the specific literature; see also Section 1.5 of the book.

Catanzaro & Naples, November 2017

Francesco Amato,
Gianmaria De Tommasi,
Alfredo Pironti

List of Acronyms

Abbreviations

DLE	differential Lyapunov equation
DLMI	differential linear matrix inequality
D/DLE	differential/difference Lyapunov equation
D/DLMI	differential/difference linear matrix inequality
FTB	finite-time boundedness
FTS	finite-time stability
IDLS	impulsive dynamical linear system
SLS	switching linear system
IO-FTS	input-output finite-time system
LMI	linear matrix inequality
LTI	linear time-invariant
LTV	linear time-varying
LS	Lyapunov stability
AS	Arbitrary switching
US	Uncertain switching

Mathematical Symbols

:	such that
$\forall$	for all
$\exists$	there exists
$:=$	equal by definition
$p \Leftrightarrow q$	p is equivalent to q
$p \Rightarrow q$	p implies q

Set Theory

$x \in A$	the element x belongs to the set A
$S_1 \bigcup S_2$	the union of the sets S_1 and S_2
$S_1 \subseteq S_2$	the set S_1 is a subset of the set S_2
$S_1 \subset S_2$	the set S_1 is a *strict* subset of the set S_2
$[t_1, t_2]$	closed interval $t_1 \leq t \leq t_2$

(t_1, t_2)	open interval $t_1 < t < t_2$
$\Omega/\mathcal{T}$	the set composed of the elements of the set Ω which do not belong to the set $\mathcal{T}$

Numerical Sets

$\mathbb{N}_0$ ($\mathbb{N}$)	nonnegative (positive) integer numbers
$\mathbb{R}$	field of real numbers
$\mathbb{R}_0^+$	nonnegative real numbers
$\mathbb{R}^n$	set of the n-tuple of real numbers
$\mathbb{R}^{m \times n}$	real matrices with m rows and n columns

Vector and Matrix Operators

x_i	the i-th element of the vector x
a_{ij}	the ij-th element of the matrix A
$\det(A)$	determinant of the square matrix A
A^{-1}	inverse of the square matrix A
A^T	transpose of matrix A
$\text{diag}(A_1, A_2, \cdots, A_r)$	block diagonal matrix with $A_1, A_2, ..., A_r$ on the diagonal
$\lambda_{max}(Q)$	maximum eigenvalue of the positive definite matrix Q
$\lambda_{min}(Q)$	minimum eigenvalue of the positive definite matrix Q
$A > 0$	A is (symmetric) positive definite
$A (\geq) 0$	A is (symmetric) positive semidefinite
$A < 0$	A is (symmetric) negative definite
$A \leq 0$	A is (symmetric) negative semidefinite
$A > B$	$A - B$ is (symmetric) positive definite
$A \geq B$	$A - B$ is (symmetric) positive semidefinite
$\text{rank}(A)$	the rank of matrix A
$A \otimes B$	Kronecker product of matrices A and B
$\langle u(\cdot), v(\cdot) \rangle$	$:= \int_\Omega u^T(\tau) v(\tau) d\tau$ where $u(\cdot) \in \mathcal{L}_p(\Omega)$, $v \in \mathcal{L}_{p'}(\Omega)$ and $1/p + 1/p' = 1$. When $p = p' = 2$ defines the internal product in $\mathcal{L}_2(\Omega)$, between $u(\cdot)$ and $v(\cdot)$.

Special Matrices

I	identity matrix; the dimension will be clear from the context
0	zero matrix; the dimension will be clear from the context

Norms

$|v|$	Euclidean norm of the vector v $\left(= (v^T v)^{1/2} = \sqrt{\sum_{i=1}^n v_i^2} \right)$
$|v|_J$	Euclidean norm of the vector v, weighted by the positive definite matrix J, defined as $(v^T J v)^{1/2}$
$|A|$	matrix norm of the matrix A induced by the Euclidean vector norm

$\|s(\cdot)\|_{p,R}$ p-norm of the vector-valued signal $s(\cdot)$, weighted by the positive definite matrix-function $R(\cdot)$, defined as $\left(\int_\Omega [s^T(\tau)R(\tau)s(\tau)]^{\frac{p}{2}}d\tau\right)^{\frac{1}{p}}$

$\|(v(\cdot),s(\cdot))\|_{2,J,R}$ 2-norm of the pair $(v,s(\cdot))$, weighted by J and $R(\cdot)$, defined as $\left(v^T J v + \int_\Omega [s^T(\tau)R(\tau)s(\tau)]d\tau\right)^{\frac{1}{2}}$

$\|s(\ )\|_p$ p norm of the vector-valued signal $s(\cdot)$, weighted by $R(\cdot) = I$

$\|(v,s(\cdot))\|_2$ 2-norm of the pair $(v,s(\cdot))$, weighted by $J = I$, and $R(\cdot) = I$

$\|s(\cdot)\|_{\infty,R}$ ∞-norm of the vector-valued signal $s(\cdot)$, weighted by $R(\cdot)$, defined as ess $\sup_{t\in\Omega}[s^T(t)R(t)s(t)]^{\frac{1}{2}}$

$\|s(\cdot)\|_\infty$ ∞-norm of the vector-valued signal $s(\cdot)$, weighted by $R(\cdot) = I$

Function Spaces

Ω the finite interval $[t_0, t_0 + T]$, $t_0 \geq 0$, $T > 0$

Ω_0 the finite interval $[0, T]$, $T > 0$

$\mathcal{L}_p(\Omega)$ the space of vector-valued signals for which the p-norm is finite on Ω

$\mathcal{L}_\infty(\Omega)$ the space of vector-valued signals for which the ∞-norm is finite on Ω

$\mathcal{W}_2$ the set of vector-valued signals $w(\cdot) \in \mathcal{L}_2(\Omega)$, with $\|w\|_{2,R} \leq 1$, and $R(\cdot)$ a positive definite matrix function of suitable dimensions

$\mathcal{W}_\infty$ the set of vector-valued signals $w(\cdot) \in \mathcal{L}_\infty(\Omega)$, with $\|w\|_{\infty,Q} \leq 1$, and $Q(\cdot)$ a positive definite matrix function of suitable dimensions.

Miscellaneous

▲   end of theorems, lemmas, corollaries and facts

△   end of examples

◇   end of assumptions, definitions, problems, procedures, exercises, remarks and proofs.

wrt   with respect to

1

Introduction

This first chapter has the twofold objective of introducing the framework of input-output finite-time stability (IO-FTS), together with the notation that will be used throughout the book, and providing some useful background on the analysis of the behavior of dynamical systems.

In order to introduce the topics dealt with in this monograph, we first recall the concept of *state* FTS, and then we will extend it to the input-output case, both with zero and nonzero initial conditions. The former extension correspond to the concept of IO-FTS, while the latter represents a generalization of the finite-time boundedness (FTB) concept, namely IO-FTS with nonzero initial conditions (IO-FTS-NZIC).

Roughly speaking, FTS involves the behavior of the system state for an autonomous dynamical system with nonzero initial conditions, while IO-FTS looks at the input-output behavior of the system, with zero initial conditions. IO-FTS-NZIC mixes the two concepts, considering the input-output finite-time control problem with a nonzero initial condition. The common points to these definitions is that they are defined over a finite-time interval and that quantitative bounds are given for the admissible signals during this interval.

1.1 Finite-Time Stability (FTS)

The concept of finite-time stability (FTS) dates back to the fifties, when it was introduced in the Russian literature ([1–3]); later, during the sixties, this concept appeared for the first time in Western journals [4–6].

Given the dynamical system

$$\dot{x}(t) = f(t, x), \quad x(t_0) = x_0, \tag{1.1}$$

where $x(t) \in \mathbb{R}^n$, we can give the following formal definition, which restates the original definition in a way consistent with the notation adopted in this monograph; in the following we consider the finite-time interval $\Omega := [t_0, t_0 + T]$, with $T > 0$.

Definition 1.1 (FTS, [2, 4, 8]) Given the time interval Ω, a set $\mathcal{X}_0 \subset \mathbb{R}^n$, and a family of sets $\mathcal{X}_t \subset \mathbb{R}^n$, system (1.1) is said to be *finite-time stable* with respect to (wrt) $(\Omega, \mathcal{X}_0, \mathcal{X}_t)$ if

$$x_0 \in \mathcal{X}_0 \Rightarrow x(t) \in \mathcal{X}_t, \quad \forall\, t \in \Omega, \tag{1.2}$$

Finite-Time Stability: An Input-Output Approach, First Edition.
Francesco Amato, Gianmaria De Tommasi, and Alfredo Pironti.

where, with a slight abuse of notation, $x(\cdot)$ denotes the solution of (1.1) starting from x_0 at time t_0.   ◇

Note that, in general, the set $\mathcal{X}_t$, called *outer (or trajectory) set*, possibly depends on time; obviously $\mathcal{X}_{t_0}$ must contain the *inner (or initial)* set $\mathcal{X}_0$, for well-posedness of Definition 1.1.

An issue that is important to clarify is why the property expressed by (1.2) is called FTS.

In order to answer this question, we recall the classical definition of Lyapunov stability (LS, [32, Ch. 4]; see also Appendix A.3). Let $\bar{x}$ be an equilibrium point for system (1.1), i.e., $f(t,\bar{x}) = 0$ for all $t \in \mathbb{R}_0^+$. The equilibrium point $\bar{x}$ is said to be *stable* in the sense of Lyapunov if for each $\varepsilon > 0$, there exists a positive scalar δ, possibly depending on t_0 and ε, such that $|x_0 - \bar{x}| < \delta(\varepsilon, t_0)$, implies

$$|x(t) - \bar{x}| < \varepsilon, \quad t \geq t_0,$$

and this holds for all $t_0 \in \mathbb{R}^+$.

The key points in the above definition are: an equilibrium point $\bar{x}$ is stable if, once an arbitrary value for ε has been fixed, which defines a *ball* centered in $\bar{x}$, then it must be possible to build an *inner* ball (of radius δ) such that, whenever the initial condition is inside such ball, the trajectory of the system starting from x_0 does not exit the *outer* ball (of radius ε). Moreover this property holds for an infinite time horizon, that is, for all t between t_0 and infinity.

Note that LS is a qualitative concept, that is, both the inner and the outer ball are not quantified; therefore, LS can be regarded as a structural property: a given equilibrium point $\bar{x}$ is either stable or it is not.

Now let us come back to Definition 1.1; even in this case there is an inner set $\mathcal{X}_0$, usually centered at an equilibrium point of system (1.1), and an outer set $\mathcal{X}_t$. FTS requires that, whenever the trajectory of (1.1) starts inside the inner set, it does not exit the outer set. From this point of view, Definition 1.1) mimics the one of LS, and this justifies the use of the term *stability*. However, differently from LS, this is only required over a *finite interval of time*, which should be possibly short with respect to steady state; i.e., FTS can be used to *shape* the behavior of the system during the transients.

Another important point is that FTS is a quantitative concept, since the inner and the outer set are specified once and for all. Therefore the same system can be finite-time stable for some choice of $\mathcal{X}_0$, $\mathcal{X}_t$, and Ω, and non-finite-time stable for a different choice of these parameters.

It is worth noting that, in principle, FTS does not necessarily requires that the inner set $\mathcal{X}_0$ contains any equilibrium point for system (1.2); however, this particular case will not be dealt with in this book, where we shall consider ellipsoidal sets centered at the origin of the state space.

A direct consequence of the discussion above is that FTS and LS are independent concepts; referring to a linear system, to simplify the terminology (see Appendix A.3), a system can be finite-time stable, despite not being stable in the sense of Lyapunov, and vice versa. While LS deals with the behavior of a system within a sufficiently long (in principle infinite) time interval, FTS is a more practical concept, useful to study the behavior of the system within a finite (possibly short) interval, and therefore it finds application whenever it is desired that the state variables do not exceed a given threshold (for example to avoid saturation or the excitation of nonlinear dynamics) during the transients.

In the following, we shall focus on linear time-varying (LTV) systems

$$\dot{x}(t) = A(t)x(t), \quad x(t_0) = x_0, \tag{1.3}$$

with $A(\cdot) : \Omega \mapsto \mathbb{R}^{n \times n}$ piecewise continuous; note that the assumption on piecewise continuity of $A(\cdot)$ guarantees existence and uniqueness of the solution of system (1.3) starting from x_0, at time t_0 (see Appendix A). Moreover, if we consider ellipsoidal state sets, i.e.,

$$\mathcal{X}_0 := \{x \in \mathbb{R}^n \text{ s.t. } x^T \Gamma_0 x \leq 1, \text{with } \Gamma_0 > 0\},$$
$$\mathcal{X}_t := \{x \in \mathbb{R}^n \text{ s.t. } x^T \Gamma(t)x < 1, \text{with } \Gamma(t) > 0 \,\forall t \in \Omega\},$$

Definition 1.1 can be rewritten as follows.

Definition 1.2 (FTS for LTV Systems [12, 33, 34])   Given the time interval Ω, a positive definite matrix Γ_0, and a continuous, positive definite matrix-valued function $\Gamma(\cdot)$ defined over Ω, such that $\Gamma(t_0) < \Gamma_0$, system (1.3) is said to be *finite-time stable* wrt $(\Omega, \Gamma_0, \Gamma(\cdot))$ if

$$x_0^T \Gamma_0 x_0 \leq 1 \Rightarrow x(t)^T \Gamma(t)x(t) < 1, \quad \forall t \in \Omega. \tag{1.4}$$

$\diamond$

As said above, the assumption that $\Gamma(t_0) < \Gamma_0$ in Definition 1.2 is needed to guarantee that the initial closed ellipsoid $\mathcal{X}_0$ is a proper subset of the open ellipsoid $\mathcal{X}_{t_0}$, hence guaranteeing the well-posedness of the definition itself.

A graphical explanation of the FTS concept is reported in Figure 1.1 for a second-order system with a constant matrix-valued function $\Gamma(t) = \Gamma$. In particular, if a system is FTS, then all the trajectories starting within the ellipse defined by Γ_0 should be like the one depicted in green in Figure 1.1. Conversely, two trajectories that are not FTS are reported in red.

In the following, we consider a numerical example.

Example 1.1 (Lyapunov stability and finite-time stability for LTI systems)   This introductory example shows the difference between the two concepts of LS and FTS for a second-order linear time-invariant (LTI) system. To this aim, let us consider the following autonomous LTI system

$$\dot{x}(t) = A_1 x(t) = \begin{pmatrix} 0 & 1 \\ -1 & -1 \end{pmatrix} x(t), \quad x(0) = x_0. \tag{1.5}$$

System (1.5) is clearly Lyapunov stable, being negative the maximum real part of the eigenvalues of the matrix A_1.

While LS is a structural property of an LTI system, FTS it is not. Indeed, given the time interval $\Omega' = [0, 2]$ if we specify the two weighting matrices in Definition 1.2 as follows

$$\Gamma'_0 = \begin{pmatrix} 10 & 0 \\ 0 & 0.9 \end{pmatrix}, \quad \Gamma' = \begin{pmatrix} 4 & 0.4 \\ 0.4 & 0.65, \end{pmatrix}$$

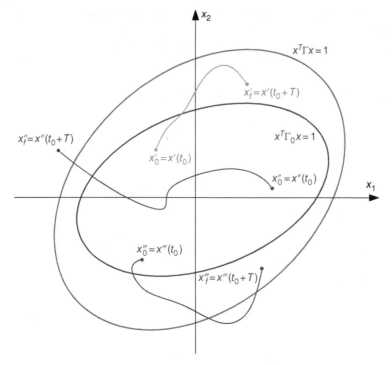

Figure 1.1 Given a time interval Ω, and the two ellipsoidal domains delimited by Γ_0 and by the constant matrix Γ, a second-order system is finite-time stable if all the trajectories over the considered time interval are like the one reported in light gray. Furthermore, in dark gray are reported two examples of trajectories that are not finite-time stable.

it turns out that system (1.5) is not FTS wrt $(\Omega', \Gamma_0', \Gamma')$, as it is clearly shown in Figure 1.2, since there is at least one state trajectory that starts within the initial domain defined by Γ_0' and that goes outside the ellipsoidal domain specified by Γ' during the time interval Ω'.

On the other hand, if we consider a different time interval for the FTS analysis, e.g., by letting $\Omega'' = [0, 0.25]$, systems (1.5) turns out to be FTS wrt $(\Omega'', \Gamma_0', \Gamma')$. Indeed, in the last case, it can be shown that all the state trajectories of (1.5) that start within the initial domain defined by Γ_0' remain within the target ellipsoidal domain defined by Γ' (one possible way to check FTS is to solve the feasibility problem reported in [35, Theorem 2.1-(v)]).

Let us now consider the following Lyapunov unstable system

$$\dot{x}(t) = A_2 x(t) = \begin{pmatrix} 0 & 1 \\ 0.9 & -0.1 \end{pmatrix} x(t), \quad x(0) = x_0. \tag{1.6}$$

Also in this case FTS is not related to LS, as it is shown in Figure 1.3, where the initial ellipsoidal domain Γ_0'' is given by

$$\Gamma_0'' = \begin{pmatrix} 0.45 & 0 \\ 0 & 1.2 \end{pmatrix},$$

Figure 1.2 Free response of the LS stable LTI system (1.5) when the initial state is set equal to $x(0) = (0 \quad 1)^T$. Although the considered LTI system is Lyapunov stable, the same system can be either FTS or not, depending on the FTS parameters.

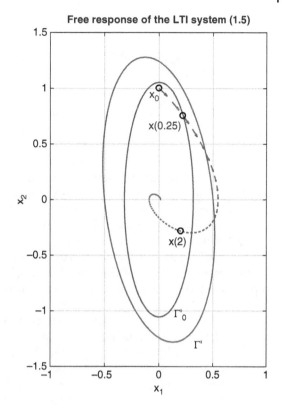

Free response of the LTI system (1.5)

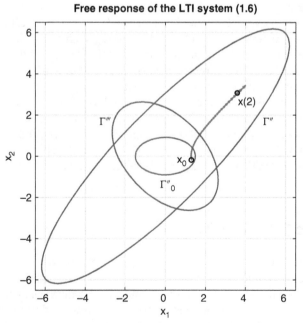

Free response of the LTI system (1.6)

Figure 1.3 Free response of the LS unstable LTI system (1.6) when the initial state is set equal to $x(0) = (1.3 - 0.2)^T$. Even when Lyapunov unstable systems are considered, the finite-time stability depends on the chosen parameters.

the finite-time interval is taken equal to $\Omega''' = [0, 2]$, and two different time-invariant target domains, defined by

$$\Gamma'' = \begin{pmatrix} 0.1070 & -0.093 \\ -0.093 & 0.107 \end{pmatrix}, \quad \Gamma''' = \begin{pmatrix} 0.18 & 0.08 \\ 0.08 & 0.18 \end{pmatrix},$$

are chosen. It is straightforward to check that system (1.6) is not FTS wrt $(\Omega''', \Gamma_0'', \Gamma''')$, while it can be proven that the same unstable system (in the Lyapunov sense) is FTS wrt $(\Omega''', \Gamma_0'', \Gamma'')$.

△

The pioneering works [1–6], although developing a nice theoretical framework, did not provide computationally tractable conditions for checking the FTS of a given dynamical system, unless simple cases were considered. Therefore, for a long period, this field of research was neglected by control scientists.

At the end of the last century, the development of the Linear Matrix Inequality theory (LMI,[7]) has fueled new interest in the field of finite-time control. In particular, FTS and stabilization have been investigated in the context of linear time-invariant systems, both continuous (e.g., [10, 13, 36–38]) and discrete-time [39, 40]. According to this modern approach to FTS, conditions for analysis and design are provided in terms of feasibility problems involving LMIs [15], by exploiting the properties of quadratic Lyapunov functions. A different approach, which looks to polyhedral Lyapunov functions, is presented in [41–44]; polyhedral Lyapunov functions are useful when the sets Γ_0 and Γ are polytopic rather than ellipsoidal. Finally, in [28, 29, 45] the concept of *annular* FTS has been introduced to take into account also a possible lower bound for the state variables.

A new impulse to the theory has been given by the use of *time-varying* quadratic Lyapunov functions, which, on the one hand has allowed to deal with the more general class of LTV systems, and more importantly, on the other hand has permitted to state non-conservative, i. e., necessary and sufficient, conditions for FTS and stabilization both for continuous an discrete-time systems [9, 11, 12, 23, 46–49]; such conditions require the solution of Differential Linear Matrix Inequalities (DLMIs, [15]), or Differential Lyapunov Equations (DLEs, [16]). In the recent papers [50, 51], the Separation Principle has been (partially) extended to the finite-time context.

An effort has been spent in order to extend the results obtained for linear systems to other contexts, such as uncertain linear systems [52], nonlinear systems [17, 18, 53], hybrid systems [19–23], and stochastic systems ([18, 23–29] among the others), and to consider mixed problems, such as FTS and pole placement [54], FTS with input constraints [55, 56], FTS and $\mathcal{H}_\infty$ control [57]. Most of the above results are collected in the monograph [35].

1.2 Input-Output Finite-Time Stability

IO-FTS represents the *natural* extension of the concept of FTS introduced in Section 1.1, to the case of non-autonomous dynamical systems.

Informally a system is said to be input-output finite-time stable if, for a given class of input signals, the output of the system does not exceed an assigned threshold during a specified time interval. As it is usual when dealing with input-output issues, the initial state of the system under consideration is assumed to be zero.

In order to formally define IO-FTS, let us consider the system

$$\dot{x}(t) = f(t, x, w), \quad x(t_0) = 0 \tag{1.7a}$$

$$y(t) = g(t, x, w), \tag{1.7b}$$

where $y(t) \in \mathbb{R}^p$ is the system output, and $w(t) \in \mathbb{R}^m$ the exogenous input, i.e., the non-manipulable input; we can give the following definition.

Definition 1.3 (IO-FTS, [30]) Given the time interval Ω, a family of sets $\mathcal{Y}_t \subset \mathbb{R}^p$, and a class of input signals $\mathcal{W}$ defined over Ω, system (1.7) is said to be *input-output finite-time stable* wrt $(\Omega, \mathcal{W}, \mathcal{Y}_t)$ if

$$w(\cdot) \in \mathcal{W} \Rightarrow y(t) \in \mathcal{Y}_t, \quad \forall t \in \Omega. \qquad \diamond$$

Similarly to what has been done between LS and *state* FTS, a parallelism can be traced between IO $\mathcal{L}_p$-stability, with particular reference to $\mathcal{L}_\infty$-stability (better known with the popular acronym of BIBO, Bounded–Input Bounded–Output, stability) and IO-FTS.

We recall that system (1.7) is said to be IO $\mathcal{L}_p$-stable [32] for any input of class $\mathcal{L}_p$ (the space of the p-integrable vector-valued functions, see Section 1.4.1), if the system exhibits a corresponding output that belongs to the same class. IO-stability of linear and nonlinear systems has been broadly studied since the early sixties [58–60]. Moreover, a number of results have been proposed in the literature to discuss robustness issues (see for example [61] and the bibliography therein).

As happened between *state* FTS and LS, also *classical* IO stability and IO-FTS differ because the latter involves signals defined over a finite-time interval and gives *quantitative* bounds on both inputs and outputs. Moreover, differently from classical IO stability, IO-FTS does not necessarily require inputs and outputs to belong to the same class of signals.

It turns out that also IO stability and IO-FTS are independent concepts. While IO stability deals with the behavior of a system within a sufficiently long (in principle infinite) time interval, IO-FTS is a more practical concept, useful to study the behavior of the system within a finite (possibly short) interval, and therefore it finds application whenever it is desired that the output variables do not exceed a given threshold during the transients, given a certain class of input signals.

Consider the non-autonomous LTV system

$$\dot{x}(t) = A(t)x(t) + F(t)w(t), \quad x(t_0) = 0 \tag{1.8a}$$

$$y(t) = C(t)x(t) + G(t)w(t), \tag{1.8b}$$

where $A(\cdot) : \Omega \mapsto \mathbb{R}^{n \times n}$, $F(\cdot) : \Omega \mapsto \mathbb{R}^{n \times m}$, $C(\cdot) : \Omega \mapsto \mathbb{R}^{p \times n}$, and $G(\cdot) : \Omega \mapsto \mathbb{R}^{p \times m}$ are piecewise continuous matrix-valued functions that describe the system dynamics. Moreover, we assume the sets in the family $\mathcal{Y}_t$ to be ellipsoidal similarly to what has been done for *state* FTS.

Given these assumptions, Definition 1.3 can be refined as follows.

Definition 1.4 (IO-FTS for LTV Systems [30, 31]) Given the time interval Ω, a class of input signals $\mathcal{W}$ defined over Ω, and a continuous, positive definite matrix-valued function $Q(\cdot)$ defined over Ω, system (1.8) is said to be input-output finite-time stable wrt $(\Omega, W, Q(\cdot))$ if

$$w(\cdot) \in W \Rightarrow y^T(t)Q(t)y(t) < 1, \forall t \in \Omega. \qquad \diamond$$

The concept of IO-FTS has been introduced by the authors in the papers [30, 62], where sufficient conditions for a given linear time-varying (LTV) system to be IO finite-time stable and stabilizable have been provided.

In [31, 63, 64] necessary and sufficient conditions for finite-time stability and stabilization of LTV systems have been proposed, while an extension of these results to the case of impulsive dynamical linear systems (a special class of hybrid systems) has been considered in [65, 66].

It should be remarked that input–output stabilization of LTV systems on a finite time horizon is tackled also in [15]. However, as for *classic* IO stability, the concept of IO stability over a finite time horizon given in [15] does not give explicit bounds on input and output signals and does not allow the input and output to belong to different classes.

Example 1.2 (BIBO stability and input-output FTS for LTI systems) Similarly to what has been discussed in Example 1.1, by means of a simple numerical example, we now show that BIBO stability and IO-FTS are two independent concepts. Let us consider the following second-order LTI system, with one exogenous input and two outputs

$$\dot{x}(t) = Ax(t) + Fw(t) = \begin{pmatrix} 0 & 1 \\ -2 & -3 \end{pmatrix} x(t) + \begin{pmatrix} 0 \\ 1 \end{pmatrix} w(t), \quad x(0) = 0 \qquad (1.9a)$$

$$y(t) = Cx(t) = \begin{pmatrix} 1 & 1 \\ 1 & -0.5 \end{pmatrix} x(t). \qquad (1.9b)$$

By simply looking at the eigenvalues of the A matrix in (1.9a), it readily follows that system (1.9) is Lyapunov stable; hence it is also BIBO stable.

We now consider the time interval $\Omega = [0, 2]$ and the class W_∞ of bounded signals over Ω, i.e., the class of signals such as $|w(t)| < 1$, $t \in \Omega$ (the reader can refer to Section 1.4.1 for a formal definition of this class of inputs).

System (1.9) can be either IO-FTS or not, depending on the choice of the weighting matrix-valued function $Q(\cdot)$ in Definition 1.2. In particular, let us consider the following two possible choices for a constant output weighting matrix

$$Q_1 = I = \begin{pmatrix} 1 & 0 \\ 0 & 1 \end{pmatrix}, \quad Q_2 = \begin{pmatrix} 2 & 1 \\ 1 & 10 \end{pmatrix}.$$

By exploiting the results that will be presented in Chapter 2 it can be shown that system (1.9) is input-output finite-time stable wrt (Ω, W_∞, Q_1), while it is not input-output finite-time stable wrt (Ω, W_∞, Q_2).

As an example of response to a bounded disturbance (exogenous input), Figure 1.4 shows the time response of system (1.9) to the unitary step, i.e.,[1] to $w(t) = \delta_{-1}(t)$. It can be noticed that, when the matrix Q_2 is considered, then the weighted output $y^T(t)Q_2y(t)$ exceeds 1, hence system (1.9) is not IO-FTS for this choice of the output weighting matrix.

---

1 Here we denote with $\delta_{-1}(t)$ the unitary Heaviside step function.

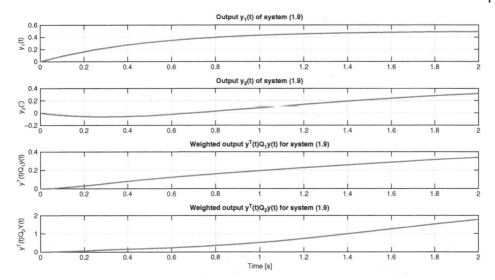

Figure 1.4 Time response of system (1.9) to the unitary step function. When the weighting matrix Q_2 is considered, then the weighted output exceeds 1; hence, the system is not IO-FTS. On the other hand, it can be proved that for all the exogenous inputs $w(t)$ belonging to the class of bounded signals in the time interval $\Omega = [0, 2]$, if the weighting matrix Q_1 is considered, then the weighted output never exceeds 1.

Similarly to what has been shown in Example 1.1, also in the case of IO-FTS it is possible to find an LTI system which is not BIBO stable but that can be either IO-FTS or not, depending on the chosen parameters. △

Combining the two concepts of *state* FTS and IO-FTS, it is possible to extend the definition of IO-FTS to the case of nonzero initial conditions (IO-FTS-NZIC). To this aim, in the following definition we consider system (1.8) when $x(t_0) = x_0 \neq 0$, i.e.,

$$\dot{x}(t) = A(t)x(t) + F(t)w(t), \quad x(t_0) = x_0 \tag{1.10a}$$
$$y(t) = C(t)x(t) + G(t)w(t), \tag{1.10b}$$

Definition 1.5 (IO-FTS-NZIC for LTV Systems [67]) Given the time interval Ω, a class of input signals $\mathcal{W}$ defined over Ω, a positive definite matrix Γ_0, and a continuous, positive definite matrix-valued function $Q(\cdot)$ defined over Ω, system (1.10) is said to be *input-output finite-time stable with nonzero initial conditions* wrt $(\Omega, \mathcal{W}, \Gamma_0, Q(\cdot))$, if

$$x_0^T \Gamma_0 x_0 \leq 1 \Rightarrow y^T(t)Q(t)y(t) < 1, \quad \forall t \in \Omega, \quad \forall w(\cdot) \in \mathcal{W}.$$

◇

It is worth to notice that the concept of IO-FTS-NZIC given in this book coincides with the definition of finite-time boundedness (FTB), when the output vector is taken equal to the state vector. The concept of FTB was introduced at the end of the last century in [8, 37, 38]; in the following years many papers have appeared in the literature dealing with the study of the FTB properties of various classes of systems (see, among others, [68–71] and the related applications [25, 72]).

1.3 FTS and Finite-Time Convergence

In the context of nonlinear systems an alternative definition of FTS has been given; essentially, this alternative definition of FTS coincides with a finite-time convergence property. For the sake of clarity, in this section we briefly discuss the main differences between the two existing finite-time frameworks.

FTS, in the sense of the convergence in finite-time of the state trajectory to an equilibrium point, is strictly related to LS and applies to autonomous systems (e.g., [73] for nonlinear continuous systems, and [74] for nonlinear impulsive systems). Hence, this alternative notion of FTS is unrelated to the notion of *state* FTS introduced in Section 1.1, since the former does not require to specify any bounding regions nor the time interval.

As in the case of the finite-time framework considered in this book, also finite-time convergence has been extended to the case of non-autonomous nonlinear systems. For example, the authors of [75] consider systems with a norm-bounded input signal over the interval $[0, +\infty]$, and a nonzero initial condition. In this case, the finite-time input-output stability is related to the property of a system to have a norm-bounded output that, after a finite time interval, does not depend anymore on the initial state. It follows that the IO-FTS considered in this book, and the extension of finite-time convergence to the input-output context are different concepts.

1.4 Background

This section introduces the notation adopted in the book, together with some useful preliminary definitions and results that will be exploited in the next chapters.

1.4.1 Vectors and signals

Let I denote the identity matrix; given a vector $v \in \mathbb{R}^n$ and a positive definite matrix $J \in \mathbb{R}^{n \times n}$, we will denote by $|v|_J$, $|v|_I =: |v|$, the Euclidean norm of v weighted by J, i.e.,

$$|v|_J = (v^T J v)^{\frac{1}{2}}.$$

Given the bounded time interval $\Omega = [t_0, t_0 + T]$. The symbol $\mathcal{L}_p(\Omega)$, denotes the space of vector-valued signals for which[2]

$$s(\cdot) \in \mathcal{L}_p(\Omega) \iff \left(\int_\Omega |s(\tau)|^p d\tau \right)^{\frac{1}{p}} < +\infty.$$

Given a symmetric positive definite, continuous matrix-valued function $R(\cdot)$, and a vector-valued signal $s(\cdot) \in \mathcal{L}_p(\Omega)$, the weighted signal norm

$$\left(\int_\Omega [s^T(\tau) R(\tau) s(\tau)]^{\frac{p}{2}} d\tau \right)^{\frac{1}{p}},$$

will be denoted by $\|s(\cdot)\|_{p,R}$. If $p = \infty$,

$$\|s(\cdot)\|_{\infty,R} = \operatorname*{ess\,sup}_{t \in \Omega} |s(t)|_{R(t)}.$$

---

2 For the sake of brevity, We denote by $\mathcal{L}_p$ the set $\mathcal{L}_p([0, +\infty))$.

When $R(\cdot) = I$, we will use the simplified notation $\|s(\cdot)\|_p$.

Given two vector-valued signals $u(\cdot) \in \mathcal{L}_p(\Omega)$, and $v(\cdot) \in \mathcal{L}_{p'}(\Omega)$, with $1/p + 1/p' = 1$, we define

$$\langle u(\cdot), v(\cdot) \rangle = \int_\Omega u^T(\tau) v(\tau) d\tau \; ; \tag{1.11}$$

when $p = p' = 2$, the operation $\langle u(\cdot), v(\cdot) \rangle$ coincides with the scalar product in $\mathcal{L}_2(\Omega)$.

Let p and p' such that $1/p + 1/p' = 1$; then the *Hölder* inequality (see [76], p. 33, [77], p. 548) states that, if $u(\cdot) \in \mathcal{L}_p(\Omega)$ and $v(\cdot) \in \mathcal{L}_{p'}(\Omega)$,

$$\int_\Omega |u^T(\tau) v(\tau)| d\tau \le \|u(\cdot)\|_p \|v(\cdot)\|_{p'}. \tag{1.12}$$

For $p = p' = 2$, the Hölder inequality is also known as *Schwarz* inequality. Moreover, the following inequality holds

$$\|u(\cdot)\|_p = \sup_{\|v\|_{p'}=1} |\langle u(\cdot), v(\cdot) \rangle|. \tag{1.13}$$

Note that, given the notation introduced for the signals and vector norms, the definitions of IO-FTS and IO-FTS-NZIC for LTV systems and ellipsoidal outputs sets can be restated as follows.

Definition 1.6 Given the time interval Ω, a class of input signals $\mathcal{W}$ defined over Ω, and a continuous, positive definite matrix-valued function $Q(\cdot)$ defined over Ω, system (1.8) is said to be *input-output finite-time stable* wrt $(\Omega, \mathcal{W}, Q(\cdot))$, if

$$w(\cdot) \in \mathcal{W} \Rightarrow \|y(\cdot)\|_{\infty,Q} < 1.$$

$\diamond$

Definition 1.7 Given the time interval Ω, a class of input signals $\mathcal{W}$ defined over Ω, a positive definite matrix Γ_0, and a continuous, positive definite matrix-valued function $Q(\cdot)$ defined over Ω, system (1.8) is said to be *input-output finite-time stable with nonzero initial conditions* wrt $(\Omega, \mathcal{W}, \Gamma_0, Q(\cdot))$, if

$$|x_0|_{\Gamma_0} \le 1 \Rightarrow \|y(\cdot)\|_{\infty,Q} < 1, \quad \forall w(\cdot) \in \mathcal{W}.$$

$\diamond$

Given the continuous, positive definite matrix $R(\cdot)$, throughout this book we will consider the following two classes of exogenous signals when dealing with IO-FTS:

i) the set of essentially bounded signals over Ω whose weighted norm is less than or equal to one

$$\mathcal{W}_\infty(\Omega, R(\cdot)) := \{ w(\cdot) \in \mathcal{L}_\infty(\Omega) : \|w\|_{\infty,R} \le 1 \},$$

ii) the set of square integrable signals over Ω whose weighted norm is less than or equal to one

$$\mathcal{W}_2(\Omega, R(\cdot)) := \{ w(\cdot) \in \mathcal{L}_2(\Omega) : \|w\|_{2,R} \le 1 \}.$$

In the rest of the book we will drop the dependency of the $\mathcal{W}$ sets on Ω and $R(\cdot)$, in order to simplify the notation.

1.4.2 Impulsive dynamical linear systems

Chapters 7–9 will deal with IO-FTS of a special class of hybrid systems, namely *Impulsive Dynamical Linear Systems (IDLSs)*. IDLSs allow to model a wide range of real-world applications whose dynamical behavior includes both time-driven and event-driven dynamics. As an example, the automatic gear-box in cruise control falls in the category of hybrid systems that can be modeled as IDLS (for more details and further examples see [78, 79]).

This section introduces the definition of this class of dynamical systems, together with some related preliminary material.

The class of IDLSs is described by the equations

$$\mathcal{ILS} : \begin{cases} \dot{x}(t) = A(t)x(t) + F(t)w(t), \quad x(t_0) = 0, \quad t \notin \mathcal{T} & \text{(1.14a)} \\ x^+(t_i) = J(t_i)x(t_i), \quad t_i \in \mathcal{T} & \text{(1.14b)} \\ y(t) = C(t)x(t) + G(t)w(t), \quad t \in \Omega, & \text{(1.14c)} \end{cases}$$

where $A(\cdot) : \Omega \mapsto \mathbb{R}^{n \times n}$, $F(\cdot) : \Omega \mapsto \mathbb{R}^{n \times m}$, $C(\cdot) : \Omega \mapsto \mathbb{R}^{p \times n}$, and $G(\cdot) : \Omega \mapsto \mathbb{R}^{p \times m}$ are piecewise continuous matrix-valued functions that describe the *continuous-time* dynamics of the system. On the other hand, $J(\cdot) : \Omega \mapsto \mathbb{R}^{n \times n}$ is the matrix-valued function that describes the *resetting law* of the system. The elements of the set $\mathcal{T} = \{t_1, t_2, \dots, t_v\} \subset \Omega$ are called *resetting times*. The finiteness of the set $\mathcal{T}$ prevents the IDLS (1.14) from exhibiting Zeno behavior; furthermore, we assume that the first resetting time $t_1 \in \mathcal{T}$ is such that $t_1 > t_0$, since we exclude the case of initial state $x(t_0) \neq 0$.

According to the continuous-time dynamics (1.14a) and the resetting law (1.14b), an IDLS presents a left-continuous trajectory with a finite jump from $x(t_i)$ to $x^+(t_i)$ at each resetting time $t_i \in \mathcal{T}$.

As we have done for LTV systems in Section A.4, we denote by $\Phi(\cdot, \cdot)$ the state transition matrix of the IDLS (1.14). It is straightforward to check the following properties for $\Phi(\cdot, \cdot)$

$$\Phi(t_0, t_0) = I, \tag{1.15a}$$

$$\frac{\partial}{\partial t} \Phi(t, t_0) = A_c(t)\Phi(t, t_0), \quad t \notin \mathcal{T} \tag{1.15b}$$

$$\Phi^+(t_i, t_0) = J(t_i)\Phi(t_i, t_0), \quad t_i \in \mathcal{T}. \tag{1.15c}$$

Furthermore, for a given t in Ω, such that $t > t_k$ and $t < t_{k+1}$, with $t_k, t_{k+1} \in \mathcal{T}$, we have

$$\Phi(t, t_0) = \Phi_{k+1}(t, t_k)J(t_k)\Phi_k(t_k, t_{k-1})J(t_{k-1}) \times \cdots$$
$$\times J(t_2)\Phi_2(t_2, t_1)J(t_1)\Phi_1(t_1, t_0), \tag{1.16}$$

where $\Phi_j(\cdot, \cdot)$ satisfies the matrix differential equation

$$\frac{\partial}{\partial t} \Phi_j(t, t_{j-1}) = A_c(t)\Phi_j(t, t_{j-1}), \quad t \in [t_{j-1}, t_j), \quad \Phi_j(t_{j-1}, t_{j-1}) = I,$$

with $j = 1, \dots, v + 1$,

Given (1.16), it is straightforward to verify that the impulsive response of the IDLS (1.14) is formally equal to (A.20), as in the case of LTV systems.

Moreover, also the Reachability Gramian of (1.14) can be recursively defined (see [80, 81] for more details), and the following lemma holds.

Lemma 1.1 ([80, 81]) The reachability Gramian $W_r(\cdot, \cdot)$ of the IDLS (1.14) is the unique positive semidefinite solution to the Difference-DLE (D/DLE)

$$\dot{W}_r(t, t_0) = A_c(t)W_r(t, t_0) + W_r(t, t_0)A^T(t) + F(t)F^T(t), \quad t \notin \mathcal{T} \tag{1.17a}$$

$$W_r^+(t_i, t_0) = J(t_i)W_r(t_i, t_0)J^T(t_i), \quad t_i \in \mathcal{T} \tag{1.17b}$$

$$W_r(t_0, t_0) = 0 \tag{1.17c}$$

▲

The definition of IO-FTS for IDLSs is the same as the one given for LTV systems; therefore, we shall refer to Definition 1.6, when considering IDLSs.

Before concluding this section, we would like to remark that IDLSs can be used to model the class of switching linear systems (SLS, [82]).

Indeed, IDLSs can also be seen as a special case of SLSs. Given a right-continuous *switching signal* σ, i.e. a piecewise constant function $\sigma(\cdot) : \mathbb{R}_0^+ \mapsto \mathcal{P} \subset \mathbb{N}$, whose discontinuities correspond to the resetting times, and the family of linear systems

$$\dot{x}(t) = A_p(t)x(t) + F_p(t)w(t), \tag{1.18a}$$

$$y(t) = C_p(t)x(t) + G_p(t)w(t), \tag{1.18b}$$

where $p \in \mathcal{P} = \{1, \dots, l\}$, the class of SLSs is given by

$$\dot{x}(t) = A_{\sigma(t)}(t)x(t) + F_{\sigma(t)}(t)w(t), \quad x(t_0) = 0, \quad t \notin \mathcal{T} \tag{1.19a}$$

$$x(t_i^+) = J(t_i)x(t_i), \quad t_i \in \mathcal{T} \tag{1.19b}$$

$$y(t) = C_{\sigma(t)}(t)x(t) + G_{\sigma(t)}w(t), \quad t \in \Omega. \tag{1.19c}$$

It follows that IDLSs (1.14) can also be seen as SLSs when the special case of a single dynamic with discontinuities in correspondence of the resetting times is considered. Hence, the two definitions are equivalent unless the linear systems in the family (1.18) have different orders.

1.5 Book Organization

After the introductory Chapter 1 (this chapter), in Chapters 2 and 3 both the analysis and design of continuous-time LTV systems in the form (1.8) are considered. We focus on the two classes of inputs, $\mathcal{W}_2$ and $\mathcal{W}_\infty$, introduced in Section 1.4.1, and some conditions guaranteeing IO-FTS and finite-time stabilization will be presented for LTV systems. More precisely, the proposed approach will lead to necessary and sufficient conditions ($\mathcal{W}_2$ case) and sufficient conditions ($\mathcal{W}_\infty$ case) for analysis and synthesis, all based on feasibility problems involving DLMIs or DLEs.

In Chapters 4 and 5 some extensions of the original definition of IO-FTS are considered. In particular, in Chapter 4 we consider the case in which the initial state is nonzero; this leads to the definition of IO-FTS-NZIC, for which some sufficient conditions are derived. In Chapter 5 we consider the usual situation where there are some amplitude constraints on the control inputs, introducing the concept of *structured* IO-FTS.

In Chapter 6 the robustness issues are considered; this represents the starting point for considering the mixed $\mathcal{H}_\infty/FTS$ control problem.

In Chapter 7, the FTS analysis for IDLS in the form (1.14) is considered; Chapter 8 deals with the design problem for the same class of systems, while in Chapter 9, the case in which the resetting times of the IDLS (1.14) are uncertain is considered.

Finally, in Chapter 10, we illustrate a hybrid architecture, where the controller is implemented by both finite-time control techniques and the classical robust control (infinite horizon) approach. This application shows that the IO-FTS approach is useful to refine the system behavior during the transients, while classical IO Lyapunov stability is a fundamental requirement to guarantee the correct behavior at steady state.

The book is equipped with five appendices. Appendix A provides some fundamental results on LTV systems; Appendix B recalls some properties of Schur Complements, which are often used in our book; Appendix C illustrates some numerical techniques to solve DLMIs and D/DLMIs, while Appendix D presents some examples of MATLAB® code used to solve optimization problems with this type of constraints. Finally, Appendix E discusses some real-world examples where the IO-FTS approach can be exploited.

There are some issues that are not investigated in this book. For example, we do not discuss the extension of the IO-FTS theory to nonlinear and/or stochastic systems (see for example [83]), systems with delays [84–86], 2D-systems [87]. IDLSs are only considered from the deterministic point of view, while there is a growing interest for impulsive and switching systems regulated by stochastic phenomena (see [88] and the bibliography therein).

For self-containedness purposes, the proofs of all the main theorems are provided; also, a reference is made to the paper where the theorem has been originally stated. Moreover each chapter is equipped with a summary, which recalls the main topics we have dealt with in the chapter itself.

All numerical computations done in the examples have been performed within the MATLAB® environment using the YALMIP parser [89] to specify the optimization problems, and by solving them either using the LMI Toolbox® [90] or other optimization solvers, such as SeDuMi [91].

2

Linear Time-Varying Systems: IO-FTS Analysis

In this chapter, by using an approach based on the Reachability Gramian, we provide some conditions to check IO-FTS of LTV systems. In particular, two necessary and sufficient conditions for IO-FTS are given when the set of $\mathcal{W}_2$ exogenous inputs is considered.

The former relies on the solution of a coupled DLMI/LMI feasibility problem; the latter is based on the existence of a suitable solution to a DLE. We show that the DLE-based condition is computationally more efficient, while the formulation via DLMI allows to solve the problem of the IO finite-time stabilization via output feedback (which is discussed in Chapter 3).

We also investigate the IO-FTS problem in presence of $\mathcal{W}_\infty$ exogenous inputs. In this case, we arrive to a sufficient condition, again consisting of a feasibility problem constrained by a coupled DLMI/LMI.

The effectiveness and computational issues are discussed in two examples; in particular, the proposed methodology is used in the second example to minimize the maximum displacement and velocity of a building subject to an earthquake of given magnitude.

Most of the material presented in this chapter is based on the content of the papers [30, 31, 62, 63].

2.1 Problem Statement

As it has already been discussed in Section 1.2, given system (1.7), if the input $w(\cdot)$ belongs to $\mathcal{L}_p$, in general it is not guaranteed that the output $y(\cdot)$ lies in the same set. For this reason it makes sense to give the definition of IO $\mathcal{L}_p$-stability. Roughly speaking (the precise definition is more involved, and the interested reader is referred to [32, Ch. 5] or to [92]), system (1.7) is said to be IO $\mathcal{L}_p$-stable, if $w(\cdot) \in \mathcal{L}_p$ implies $y(\cdot) \in \mathcal{L}_p$. In the context of $\mathcal{L}_p$-stability, the most popular cases are the ones with $p = 2$ and $p = \infty$.

While the concept of $\mathcal{L}_p$-stability is generally referred to an infinite interval of time, in this book we are interested to study the input-output behavior of the system over a finite time interval.

To this end, it is important to recall that system (1.8) can be also viewed as a linear *operator* that maps the (exogenous) input signals $w(\cdot)$ into the output signals $y(\cdot)$.

As for the *type* of input signals $w(\cdot)$ to be considered in the definition of IO-FTS, this book deals with the two classes of signals defined over Ω introduced in Section 1.4.1,

Finite-Time Stability: An Input-Output Approach, First Edition.
Francesco Amato, Gianmaria De Tommasi, and Alfredo Pironti.
© 2018 John Wiley & Sons Ltd. Published 2018 by John Wiley & Sons Ltd.

namely the set of essentially bounded signals $\mathcal{W}_\infty$, and the square integrable signals $\mathcal{W}_2$. Given these two classes of exogenous inputs, Section 2.2 provides two necessary and sufficient conditions to check IO-FTS of system (1.8) for $\mathcal{W}_2$ inputs, while in Section 2.3 a sufficient condition for IO-FTS of system (1.8) in presence of $\mathcal{W}_\infty$ inputs is proven.

2.2 IO-FTS for $\mathcal{W}_2$ Exogenous Inputs

This section discusses the IO-FTS of LTV systems when the exogenous input belongs to the $\mathcal{W}_2$ class. In particular, in Section 2.2.2 we will state a couple of necessary and sufficient conditions to check IO-FTS of system (1.8), when the family of square integrable exogenous inputs is considered.

To do this, we need to consider the case when the feedthrough matrix $G(\cdot)$ in (1.8) is equal to zero.

If this assumption were not true, the concept of IO-FTS wrt $\mathcal{W}_2$ would be ill posed. Indeed, it is straightforward to recognize that $\mathcal{W}_2$ includes signals that are unbounded on an interval of zero measure included in Ω. When $G(\cdot) \neq 0$, in presence of such signals, it would exists at least one time instant where the output $y(\cdot)$ would be unbounded. Hence, when $G(\cdot) \neq 0$, system (1.8) cannot be IO-FTS with respect to $\mathcal{W}_2$.

For this reason we set $G(\cdot) = 0$ in Ω, and in this section we will deal with the following LTV system

$$\mathcal{LS} : \begin{cases} \dot{x}(t) = A(t)x(t) + F(t)w(t), & x(t_0) = 0 \\ y(t) = C(t)x(t). \end{cases} \tag{2.1}$$

2.2.1 Preliminaries

In order to prove the main results of Section 2.2.2, we first state an equivalent definition of IO-FTS that can be easily derived when the LTV system (2.1) is regarded as a linear operator that maps signals from the space $\mathcal{W}_2$ to the space $\mathcal{W}_\infty$, i.e.:

$$\mathcal{LS} : w(\cdot) \in \mathcal{L}_2(\Omega) \mapsto y(\cdot) \in \mathcal{L}_\infty(\Omega). \tag{2.2}$$

Moreover, by equipping the $\mathcal{L}_2(\Omega)$ and $\mathcal{L}_\infty(\Omega)$ spaces with the weighted norms $\| \cdot \|_{2,R}$ and $\| \cdot \|_{\infty,Q}$ (see Section 1.4.1), then the induced norm of the linear operator (2.2) is given by

$$\|\mathcal{LS}\|_{R,Q} = \sup_{\|w(\cdot)\|_{2,R}=1} \|y(\cdot)\|_{\infty,Q},$$

that is, the norm of $\mathcal{LS}$ is computed considering the input signals in $\mathcal{W}_2$.

Given this interpretation of system (2.1) as a linear operator, requiring the IO-FTS wrt $(\Omega, \mathcal{W}_2, Q(\cdot))$ is equivalent to require that $\|\mathcal{LS}\|_{R,Q} < 1$; therefore, the following theorem holds.

Theorem 2.1 Given the time interval Ω, the class of input signals $\mathcal{W}_2$, and a continuous, positive definite matrix-valued function $Q(\cdot)$ defined over Ω, system (2.1) is IO-FTS wrt $(\Omega, \mathcal{W}_2, Q(\cdot))$ if and only if $\|\mathcal{LS}\|_{R,Q} < 1$. ▲

Given the linear operator (2.2), its *dual operator*, defined as

$$\overline{\mathcal{LS}} : z(\cdot) \in \mathcal{L}_1(\Omega) \mapsto v(\cdot) \in \mathcal{L}_2(\Omega),$$

corresponds to the system dual to (1.8) (see also [93, p. 236])

$$\overline{\mathcal{LS}} : \begin{cases} \dot{\tilde{x}}(t) = -A^T(t)\tilde{x}(t) - C^T(t)z(t) \\ v(t) = F^T(t)\tilde{x}(t) \end{cases}.$$ (2.3)

The norm of $\overline{\mathcal{LS}}$ is defined as

$$\|\overline{\mathcal{LS}}\|_{Q,R} - \sup_{\|z(\cdot)\|_{1,Q}=1} [\|v(\cdot)\|_{2,R}].$$

Moreover, by definition of dual operator ([76]), given $z(\cdot) \in \mathcal{L}_1(\Omega)$ and $w(\cdot) \in \mathcal{L}_2(\Omega)$, we have that

$$\langle z, \mathcal{LS}w \rangle = \langle \overline{\mathcal{LS}}z, w \rangle.$$ (2.4)

Therefore (2.4) reads

$$\langle z, \mathcal{LS}w \rangle = \int_\Omega z^T(t) \int_\Omega H(t,\tau)w(\tau)d\tau dt = \int_\Omega \left(\int_\Omega z^T(t)H(t,\tau)dt \right) w(\tau)d\tau$$
$$= \int_\Omega \left(\int_\Omega z^T(t)\overline{H}^T(\tau,t)dt \right) w(\tau)d\tau = \langle \overline{\mathcal{LS}}z, w \rangle,$$

where, being $G(\cdot) = 0$, the impulse response of (2.1) is equal to

$$H(t,\tau) = C(t)\Phi(t,\tau)F(\tau)\delta_{-1}(t-\tau),$$ (2.5)

while

$$\overline{H}(t,\tau) = H^T(\tau,t) = F^T(t)\Phi^T(\tau,t)C^T(\tau)\delta_{-1}(\tau-t),$$ (2.6)

is the impulse response of the dual system (2.3).

Furthermore, by following the guidelines of [94], it can be shown that the norm of the operator $\mathcal{LS}$ equals the norm of its dual.

Lemma 2.1 Given the operators $\mathcal{LS}$ and $\overline{\mathcal{LS}}$, the following holds true

$$\|\mathcal{LS}\|_{R,Q} = \|\overline{\mathcal{LS}}\|_{Q,R}.$$ (2.7)

▲

Proof: At a first stage, let us assume that $R(\cdot) = Q(\cdot) = I$, let $\|\mathcal{LS}\| := \|\mathcal{LS}\|_{I,I}$, and $\|\overline{\mathcal{LS}}\| := \|\overline{\mathcal{LS}}\|_{I,I}$; then, for a given $z(\cdot) \in \mathcal{L}_1(\Omega)$, $\|z(\cdot)\|_1 = 1$, we have that

$$\|\overline{\mathcal{LS}}z\|_2 = \sup_{\|v(\cdot)\|_2=1} |\langle \overline{\mathcal{LS}}z, v \rangle| \qquad \text{in view of (1.13) with } p = p' = 2$$

$$= \sup_{\|v(\cdot)\|_2=1} |\langle z, \mathcal{LS}v \rangle|$$

$$= \sup_{\|v(\cdot)\|_2=1} |\int_\Omega z^T(t) \mathcal{LS}v(t)dt|$$

$$\leq \sup_{\|v(\cdot)\|_2=1} \int_\Omega |z^T(t) \mathcal{LS}v(t)|dt$$

$$\leq \|z(\cdot)\|_1 \sup_{\|v(\cdot)\|_2=1} \|\mathcal{LS}v\|_\infty \quad \text{in view of (1.12) with } p = 1, \ p' = \infty$$

$$= \|z(\cdot)\|_1 \|\mathcal{LS}\|.$$ (2.8)

Therefore we have that $\|\overline{\mathcal{LS}}\| \leq \|\mathcal{LS}\|$.

Conversely, by exploiting (1.13) with $p = \infty$ and $p' = 1$, we have

$$\|\mathcal{L}Sw\|_\infty = \sup_{\|y\|_1=1} |\langle \mathcal{L}Sw, y\rangle|$$

$$= \sup_{\|y\|_1=1} |\langle w, \overline{\mathcal{L}S}y\rangle|$$

$$\leq \|w(\cdot)\|_2 \sup_{\|y(\cdot)\|_1=1} \|\overline{\mathcal{L}S}y\|_2, \tag{2.9}$$

where we have used the Schwarz inequality, i.e. (1.12) with $p = p' = 2$.

Therefore we have that $\|\mathcal{L}S\| \leq \|\overline{\mathcal{L}S}\|$; from this the proof follows.

When the weighting matrices do not equal the identity, we can apply the proof outlined above by modifying the matrices of the LTV system (2.1) as follows

$$\tilde{F}(t) = F(t)R(t)^{-\frac{1}{2}}, \quad \tilde{C}(t) = Q^{\frac{1}{2}}(t)C(t). \tag{2.10}$$

$$\Diamond$$

The next lemma is a generalization of a result given in [95] relatively to the case of LTV systems, and it allows us to compute the norm of the $\mathcal{L}S$ operator as a function of the spectral radius of the Reachability Gramian defined in Section A.4. In order to prove this result, the following lemma is needed.

Lemma 2.2 If

$$v(t) \triangleq \int_\Omega f(t, \tau) d\tau, \quad t \in \Omega,$$

with $\|f(\cdot, \tau)\|_2$ integrable, then the following inequality hold

$$\|v(\cdot)\|_2 \leq \int_\Omega \|f(\cdot, \tau)\|_2 \, d\tau. \tag{2.11}$$

▲

Proof: Let us first prove that, given $k > 0$ and a signal $v(\cdot)$, the inequality

$$\|v(\cdot)\|_2 \leq k, \tag{2.12}$$

holds *if and only if*, for all $z(\cdot)$, with $\|z(\cdot)\|_2 \leq 1$,

$$\langle v(\cdot), z(\cdot)\rangle \leq k. \tag{2.13}$$

(only if). Let $\|v(\cdot)\|_2 \leq k$ and assume, *ad absurdum*, that there exists a $z(\cdot)$ such that

$$\|z(\cdot)\|_2 \leq 1,$$

and

$$\langle v(\cdot), z(\cdot)\rangle > k.$$

Since, from the Schwarz inequality , it follows that

$$\langle v(\cdot), z(\cdot)\rangle \leq |\langle v(\cdot), z(\cdot)\rangle|$$

$$\leq \int_\Omega |v^T(t)z(t)| dt$$

$$\leq \|v(\cdot)\|_2 \cdot \|z(\cdot)\|_2, \tag{2.14}$$

the initial assumption is contradicted.

(*if*). Let $\langle v(\cdot), z(\cdot) \rangle \leq k$, for all $z(\cdot)$ such that $\|z(\cdot)\|_2 \leq 1$, and assume, *ad absurdum*, that $\|v(\cdot)\|_2 > k$. By picking

$$z(t) := \frac{1}{\|v(\cdot)\|_2} v(t),$$

it turns out that $\|z(\cdot)\|_2 = 1$, and

$$\langle v(\cdot), z(\cdot) \rangle = \frac{1}{\|v(\cdot)\|_2} \int_\Omega v^T(t) v(t) dt = \|v(\cdot)\|_2 > k,$$

which, again, contradicts the initial assumption.

It is now possible to exploit the equivalence between (2.12) and (2.13) in order to prove Lemma 2.2. Let $z(\cdot)$ be a signal such that $\|z(\cdot)\|_2 \leq 1$, then

$$\langle v(\cdot), z(\cdot) \rangle = \int_\Omega \left(\int_\Omega f^T(t, \tau) d\tau \right) z(t) dt$$

$$= \int_\Omega \int_\Omega f^T(t, \tau) z(t) dt d\tau \tag{2.15a}$$

$$\leq \int_\Omega \left| \int_\Omega f^T(t, \tau) z(t) dt \right| d\tau$$

$$\leq \int_\Omega \int_\Omega |f^T(t, \tau) z(t)| dt d\tau$$

$$\leq \int_\Omega \|f(\cdot, \tau)\|_2 \cdot \|z(\cdot)\|_2 d\tau, \tag{2.15b}$$

$$\leq \int_\Omega \|f(\cdot, \tau)\|_2 d\tau,$$

where equality (2.15a) follows from the Fubini's theorem [76, p. 18], while (2.15b) follows from the Schwarz inequality.

Letting $k = \int_\Omega \|f(\cdot, \tau)\| d\tau$, inequality (2.11) follows from the equivalence between (2.12) and (2.13). $\diamondsuit$

Lemma 2.2 can now be exploited to prove the following theorem.

Theorem 2.2 ([31, 63]) Given the LTV system (2.1), the norm of the corresponding linear operator (2.2) is given by

$$\|\mathcal{L S}\|_{R,Q} = \sup_{t \in \Omega} \lambda_{\max}^{\frac{1}{2}} \left(Q^{\frac{1}{2}}(t) C(t) W(t_0, t) C^T(t) Q^{\frac{1}{2}}(t) \right), \tag{2.16}$$

where $\lambda_{\max}(\cdot)$ denotes the maximum eigenvalue of the argument, and $W(t, t_0)$ is the positive semidefinite matrix-valued function solution of

$$\dot{W}(t, t_0) = A(t) W(t, t_0) + W(t, t_0) A^T(t) + F(t) R(t)^{-1} F^T(t) \tag{2.17a}$$

$$W(t_0, t_0) = 0 \tag{2.17b}$$

▲

Proof: For the sake of simplicity, we consider the weighting matrices equal to the identity; we will discuss how to take into account these matrices at the end of the proof. Hence, letting, for all $t \in \Omega$,

$$R(t) = Q(t) = I,$$

it follows that the solution of (2.17) is given by the Reachability Gramian $W_r(t, t_0)$ (see Definition A.4 in Section A.4).

First note that, in view of (2.7), proving (2.16), with unitary weights, is equivalent to show the same equality by replacing the $\mathcal{LS}$ operator with its dual one, that is

$$\|\overline{\mathcal{LS}}\| = \sup_{t \in \Omega} \lambda_{\max}^{\frac{1}{2}}(C(t)W_r(t, t_0)C^T(t)).$$

Taking into account the definition of $\overline{\mathcal{LS}}$, letting

$$\Upsilon(t) = \int_\Omega H(t, \sigma)H^T(t, \sigma)d\sigma, \tag{2.18}$$

and denoting by $v(\cdot)$ the output of system (2.3), we obtain

$$\|v(\cdot)\|_2 = \|\int_\Omega \overline{H}(\cdot, \tau)z(\tau)d\tau\|_2$$

$$\leq \int_\Omega \|\overline{H}(\cdot, \tau)z(\tau)\|_2 d\tau \qquad \text{by Lemma 3,}$$

$$= \int_\Omega \left(z^T(\tau)\int_\Omega \overline{H}^T(t, \tau)\overline{H}(t, \tau)dt\, z(\tau)\right)^{\frac{1}{2}} d\tau$$

$$= \int_\Omega (z^T(\tau)\Upsilon(\tau)z(\tau))^{\frac{1}{2}} d\tau \qquad \text{by (2.6) and (2.18)}$$

$$= \int_\Omega |\Upsilon(\tau)^{\frac{1}{2}}z(\tau)| d\tau$$

$$\leq \int_\Omega |\Upsilon(\tau)^{\frac{1}{2}}| \cdot |z(\tau)| d\tau$$

$$= \int_\Omega \lambda_{\max}^{\frac{1}{2}}(\Upsilon(\tau)) \cdot |z(\tau)|\, d\tau \qquad \text{since } \Upsilon(t) \geq 0, \quad t \in \Omega$$

$$\leq \sup_{t \in \Omega} \lambda_{\max}^{\frac{1}{2}}(\Upsilon(t)) \cdot \int_\Omega |z(\tau)|\, d\tau$$

$$= \sup_{t \in \Omega} \lambda_{\max}^{\frac{1}{2}}(\Upsilon(t)) \cdot \|z(\cdot)\|_1,$$

thus

$$\|\overline{\mathcal{LS}}\| \leq \sup_{t \in \Omega} \lambda_{\max}^{\frac{1}{2}}(\Upsilon(t)). \tag{2.19}$$

From Definition 14 the matrix-valued function $\Upsilon(t)$ is equal to

$$\Upsilon(t) = C(t)W_r(t, t_0)C^T(t);$$

hence (2.19) is equivalent to

$$\|\overline{\mathcal{LS}}\| \leq \sup_{t \in \Omega} \lambda_{\max}^{\frac{1}{2}}(C(t)W_r(t, t_0)C^T(t)). \tag{2.20}$$

The last part of the proof is devoted to show that (2.20) is actually an equality. In order to do that, let us denote by γ the right-hand side in equation (2.20). Hence, (2.20) can be rewritten as

$$\|\overline{\mathcal{L}S}\| \leq \gamma. \tag{2.21}$$

We shall now build a sequence of inputs to system (2.3) with unit norm in $\mathcal{L}_1(\Omega)$, such that the sequence of the norms of the corresponding output signals converges to γ.

To this end consider a subset $\Omega' \subset \Omega$, such that, for all $t \in \Omega'$,

$$\lambda_{\max}^{\frac{1}{2}}(C(t)W_r(t,t_0)C^T(t)) \geq \gamma - \varepsilon,$$

with $\varepsilon > 0$. Now let $\sigma \in \Omega'$, and consider the sequence of inputs

$$z_{\varepsilon,\alpha}(t) = h(\sigma)u_\alpha(t),$$

where $h(\sigma)$ is the unit norm eigenvector corresponding to the maximum eigenvalue of $C(\sigma)W_r(\sigma,t_0)C^T(\sigma)$, and u_α is a sequence of positive scalar functions with unit norm in $\mathcal{L}_1(\Omega)$, which approach the Dirac delta function applied in σ as $\alpha \mapsto 0$. Let

$$v_{\varepsilon,\alpha}(t) = \overline{\mathcal{L}S}z_{\varepsilon,\alpha}(t) = \int_\Omega \overline{H}(t,\tau)z_{\varepsilon,\alpha}(\tau)d\tau.$$

It is simple to recognize that, as $\alpha \to 0$, we have

$$v_{\varepsilon,\alpha}(\cdot) \to \int_\Omega \overline{H}(t,\tau)h(\sigma)\delta(\tau-\sigma)d\tau = \overline{H}(t,\sigma)h(\sigma) \quad \text{in } \mathcal{L}_2(\Omega).$$

Therefore

$$\lim_{\alpha\to0}\|v_{\varepsilon,\alpha}(\cdot)\|_2^2 = \int_\Omega h^T(\sigma)\overline{H}^T(t,\sigma)\overline{H}(t,\sigma)h(\sigma)dt$$

$$= h^T(\sigma)\int_\Omega H(\sigma,t)H^T(\sigma,t)dt\, h(\sigma) \qquad \text{by (2.6)}$$

$$= h^T(\sigma)C(\sigma)W_r(\sigma,t_0)C^T(\sigma)h(\sigma).$$

Hence, we can conclude that

$$\lim_{\alpha\to0}\|v_{\varepsilon,\alpha}(\cdot)\|_2 = \lambda_{\max}^{\frac{1}{2}}(C(\sigma)W_r(\sigma,t_0)C^T(\sigma)) \geq \gamma - \varepsilon,$$

since $\sigma \in \Omega'$; therefore, given $\eta > 0$, it is possible to choose a sufficiently small α such that

$$\|v_{\varepsilon,\alpha}(\cdot)\|_2 \geq \gamma - \varepsilon - \eta.$$

Taking into account (2.21), that the scalars ε and η can be chosen arbitrarily small, and that the set of the signals $z_{\varepsilon,\alpha}$ is a subset of the set of the unit norm signals in $\mathcal{L}_1(\Omega)$, we can conclude that

$$\gamma \geq \|\overline{\mathcal{L}S}\| = \sup_{\|z(\cdot)\|_1=1} \|v(\cdot)\|_2$$

$$\geq \sup_{z_{\varepsilon,\alpha}(\cdot)} \|v_{\varepsilon,\alpha}(\cdot)\|_2 = \gamma$$

From the last chain of inequality the proof follows.

In order to conclude the proof, we now need to explicitly include the weighting matrices $R(\cdot)$ and $Q(\cdot)$. In fact, it is straightforward to extend the proof to the case $R(t) \neq I$, and $Q(\cdot) \neq I$, by modifying the matrices of the LTV system (2.1) according to (2.10), and replacing $W_r(t, t_0)$ by $W(t, t_0)$. ◇

We conclude the section with the following technical lemma, which will be useful to prove the main result of the chapter in Section 2.2.2.

Lemma 2.3 Given $\epsilon > 0$, the solution of the DLE

$$\dot{W}_\epsilon(t, t_0) = A(t)W_\epsilon(t, t_0) + W_\epsilon(t, t_0)A^T(t) + F(t)R(t)^{-1}F^T(t) + \epsilon I, \tag{2.22a}$$

$$W_\epsilon(t_0, t_0) = \epsilon I \tag{2.22b}$$

is the positive definite matrix function

$$W_\epsilon(t, t_0) = W(t, t_0) + \epsilon \Phi(t, t_0)\Phi^T(t, t_0) + \epsilon \int_{t_0}^{t} \Phi(t, \tau)\Phi^T(t, \tau)d\tau, \tag{2.23}$$

where $W(\cdot, \cdot)$ is the solution of equations (2.17). ▲

Proof: The proof follows from direct substitution of $W_\epsilon(\cdot, \cdot)$ in (2.22), and by the fact that the matrix $\Phi(t, t_0)\Phi^T(t, t_0)$ is positive definite. ◇

2.2.2 Necessary and sufficient conditions for IO-FTS for $\mathcal{W}_2$ exogenous inputs

Given the preliminary results introduced in Section 2.2.1, we are now ready to state two necessary and sufficient conditions for the IO-FTS of system (2.1).

Theorem 2.3 (Necessary and sufficient conditions for IO-FTS, $\mathcal{W}_2$ inputs [31]) Given the time interval Ω, the class of inputs $\mathcal{W}_2$, a continuous, positive definite matrix-valued function $Q(\cdot)$, the following statements are equivalent:

i) System (2.1) is IO-FTS wrt $(\Omega, \mathcal{W}_2, Q(\cdot))$.
ii) The inequality

$$\lambda_{\max}\left(Q^{\frac{1}{2}}(t)C(t)W(t, t_0)C^T(t)Q^{\frac{1}{2}}(t)\right) < 1 \tag{2.24}$$

holds for all $t \in \Omega$, where $W(\cdot, \cdot)$ is the positive semidefinite solution of the DLE (2.17).
iii) The coupled DLMI/LMI

$$\begin{pmatrix} \dot{P}(t) + A^T(t)P(t) + P(t)A(t) & P(t)F(t) \\ F^T(t)P(t) & -R(t) \end{pmatrix} < 0 \tag{2.25a}$$

$$P(t) > C^T(t)Q(t)C(t), \tag{2.25b}$$

admits a piecewise continuously differentiable, positive definite solution $P(\cdot)$ over Ω. ▲

Proof: We will prove the equivalence of the three statements by showing that **i)** ⇒ **ii)**, **ii)** ⇒ **iii)**, and **iii)** ⇒ **i)**.

[**i)** ⇒ **ii)**]. The proof readily follows from Theorems 2.1 and 2.2.

[**ii)** ⇒ **iii)**]. Given $\epsilon > 0$, consider the DLE (2.22), whose solution $W_\epsilon(\cdot, \cdot)$, given by (2.23), is positive definite and satisfies the DLMI

$$-\dot{W}_\epsilon(t, t_0) + A(t)W_\epsilon(t, t_0) + W_\epsilon(t, t_0)A^T(t) + F(t)R(t)^{-1}F^T(t) < 0. \tag{2.26}$$

Now letting

$$W_\epsilon(t, t_0) = P^{-1}(t),$$

it follows that $\dot{W}_\epsilon(t, t_0) = -P^{-1}(t)\dot{P}(t)P^{-1}(t)$, and inequality (2.26) can be written as

$$P^{-1}(t)\dot{P}(t)P^{-1}(t) + A(t)P^{-1}(t) + P^{-1}(t)A^T(t) + F(t)R^{-1}(t)F^T(t) < 0, \tag{2.27}$$

for all $t \in \Omega$. By pre- and post-multiplying (2.27) by $P(t)$ we obtain

$$\dot{P}(t) + P(t)A(t) + A^T(t)P(t) + P(t)F(t)R^{-1}(t)F^T(t)P(t) < 0, \tag{2.28}$$

and (2.25a) readily follows by applying Schur complements.

In order to prove (2.25b), first note that $W_\epsilon(\cdot, \cdot) \xrightarrow{\epsilon \to 0} W(\cdot, \cdot)$, hence, by continuity arguments, there exists a sufficiently small ϵ such that

$$\lambda_{\max}\left(Q^{\frac{1}{2}}(t)C(t)W_\epsilon(t, t_0)C^T(t)Q^{\frac{1}{2}}(t)\right) < 1. \tag{2.29}$$

Moreover, condition (2.29) is equivalent to

$$I - Q^{\frac{1}{2}}(t)C(t)P^{-1}(t)C^T(t)Q^{\frac{1}{2}}(t) > 0, \tag{2.30}$$

that, by applying Schur complements, reads

$$\begin{pmatrix} I & Q^{\frac{1}{2}}(t)C(t) \\ C^T(t)Q^{\frac{1}{2}}(t) & P(t) \end{pmatrix} > 0. \tag{2.31}$$

From [96, Lemma 5.3] inequality (2.31) is equivalent to

$$\begin{pmatrix} P(t) & C^T(t)Q^{\frac{1}{2}}(t) \\ Q^{\frac{1}{2}}(t)C(t) & I \end{pmatrix} > 0,$$

which yields (2.25b) by applying again Schur complements.

[**iii)** ⇒ **i)**]. We have already shown that, by applying Schur complements, condition (2.25a) is equivalent to (2.28). Now, let us consider the quadratic function $x^T(t)P(t)x(t)$; the derivative with respect to time is given by

$$\frac{d}{dt}(x^T(t)P(t)x(t)) = x^T(t)\dot{P}(t)x(t) + \dot{x}^T(t)P(t)x(t) + x^T(t)P(t)\dot{x}(t)$$
$$= x^T(t)(\dot{P}(t) + A^T(t)P(t) + P(t)A(t))x(t) + w^T(t)F^T(t)P(t)x(t)$$
$$+ x^T(t)P(t)F(t)w(t).$$

Thus condition (2.28) implies that

$$\frac{d}{dt}(x^T(t)P(t)x(t)) < w^T(t)F^T(t)P(t)x(t) + x^T(t)P(t)F(t)w(t)$$
$$- x^T(t)P(t)F(t)R^{-1}(t)F^T(t)P(t)x(t).$$

Let $v(t) = (R^{1/2}(t)w(t) - R^{-1/2}(t)F^T(t)P(t)x(t))$, then

$$v^T(t)v(t) = w^T(t)R(t)w(t) + x^T(t)P(t)F(t)R^{-1}(t)F^T(t)P(t)x(t)$$
$$- w^T(t)F^T(t)P(t)x(t) - x^T(t)P(t)F(t)w(t).$$

It follows that

$$\frac{d}{dt}(x^T(t)P(t)x(t)) < w^T(t)R(t)w(t) - v^T(t)v(t) \leq w^T(t)R(t)w(t). \tag{2.32}$$

Integrating (2.32) between t_0 and $t \in \Omega$, taking into account that $x(t_0) = 0$, and that $w(\cdot)$ belongs to $\mathcal{W}_2$, we obtain

$$x^T(t)P(t)x(t) \leq \int_{t_0}^{t} w^T(\sigma)R(\sigma)w(\sigma)d\sigma \leq \|w\|_{2,R}^2 \leq 1.$$

By exploiting condition (2.25b), it follows that

$$y^T(t)Q(t)y(t) = x^T(t)C^T(t)Q(t)C(t)x(t) < x^T(t)P(t)x(t) \leq 1,$$

for all $t \in \Omega$; hence system (2.1) is IO-FTS wrt $(\Omega, \mathcal{W}_2, Q(\cdot))$. ◇

As it will be shown in Section 2.2.3, the necessary and sufficient condition based on the Reachability Gramian turns out to be computationally more efficient than the coupled DLMI/LMI when checking IO-FTS. However, the coupled DLMI/LMI can be effectively used to design a finite-time stabilizing controller, as it will be shown in Chapter 3.

The following two corollaries deal with the special case in which the linear system (2.1) is time-invariant and the weighting matrices R and Q are constant. In this special case, dealing with the time interval $\Omega := [t_0, t_0 + T]$, is equivalent to deal with the time interval $\Omega_0 := [0, T]$; it follows that conditions (2.24) and (2.25b) in Theorem 2.3 need to be checked only for $t = T$.

Corollary 2.1 Given the time interval Ω_0, and two positive definite matrices R and Q, the LTI system

$$\dot{x}(t) = Ax(t) + Fw(t), \quad x(0) = 0 \tag{2.33a}$$

$$y(t) = Cx(t), \tag{2.33b}$$

is IO-FTS wrt $(\Omega_0, \mathcal{W}_2, Q)$ *if and only if*

$$\lambda_{\max}\left(Q^{\frac{1}{2}}CW(T)C^TQ^{\frac{1}{2}}\right) < 1, \tag{2.34}$$

where $W(\cdot)$ is the positive semidefinite solution of the DLE

$$-\dot{W}(t) + AW(t) + W(t)A^T + FR^{-1}F^T = 0, \quad W(0) = 0.$$

▲

Proof: The proof readily follows from Theorem 2.3, taking into account the *monotonicity* of the Reachability Gramian in the LTI case (see Remark A.1). Indeed, if condition (2.34) is satisfied, then

$$\lambda_{\max}\left(Q^{\frac{1}{2}}CW(t)C^TQ^{\frac{1}{2}}\right) < 1, \quad \forall t \in \Omega_0.$$

Corollary 2.1 can be exploited to prove the following result.

Corollary 2.2 Given the time interval Ω_0, and two positive definite matrices R and Q, system (2.33) is IO-FTS wrt (Ω_0, W_2, Q) *if and only if* the coupled DLMI/LMI

$$\begin{pmatrix} \dot{P}(t) + A^T P(t) + P(t)A & P(t)F \\ F^T P(t) & -R \end{pmatrix} < 0, \qquad \forall t \in \Omega_0 \tag{2.35a}$$

$$P(T) > C^T QC, \tag{2.35b}$$

admits a piecewise continuously differentiable, positive definite solution $P(\cdot)$ over Ω_0. ▲

Remark 2.1 Note that, even when the system is time-invariant, the solution of a DLMI is required in order to check IO-FTS of the given system. This is due to the finite time nature of the problem we are dealing with (see also the optimal control problem defined over a finite horizon [97]). To this regard, see also the discussion in Section 2.4. ◊

2.2.3 Computational issues

This section shows the effectiveness of the results proposed in Section 2.2.2 by means of an example. In particular, some computational issues are discussed, by comparing the numerical efficiency of the two necessary and sufficient conditions stated in Theorem 2.3.

Example 2.1 Let us consider the LTV system

$$\dot{x}(t) = \begin{pmatrix} 0.5 + t & 0.1 \\ 0.4 & -0.3 + t \end{pmatrix} x(t) + \begin{pmatrix} 1 \\ 1 \end{pmatrix} w(t) \tag{2.36a}$$

$$y(t) = \begin{pmatrix} 1 & 1 \end{pmatrix}, \tag{2.36b}$$

together with the following IO-FTS parameters

$$R = 1, \quad \Omega = [0, 0.5].$$

The conditions stated in Theorem 2.3 are, in principle, necessary and sufficient. However, due to the time-varying nature of the involved matrices, the numerical implementation of such conditions introduces some unavoidable conservativeness.

In order to compare the conditions stated in Theorem 2.3, the output weighting matrix is left as a free parameter. More precisely, we introduce the parameter q_{max}, defined as the maximum value of the constant q such that system (2.36) is IO finite-time stable wrt (Ω, W_2, q), and use the conditions stated in Theorem 2.3 to obtain an estimate of q_{max}.

In order to recast the DLMI condition (2.25) in terms of LMIs, the matrix-valued function $P(\cdot)$ has been assumed continuous and piecewise affine, according to the procedure described in Appendix C.1. In particular, the time interval Ω has been divided into $n = T/T_s$ subintervals, and the time derivatives of $P(\cdot)$ have been considered constant in each subinterval, guaranteeing the continuity of $P(\cdot)$ at the extrema of each subinterval.

Since the equivalence between IO-FTS and condition (2.25) holds when $T_s \to 0$, the maximum value of Q satisfying condition (2.25), namely q_{max}, has been evaluated for

Table 2.1 Maximum values of q satisfying Theorem 2.3 for the LTV system (2.36).

IO-FTS condition	Sample Time (T_s)	Estimate of q_{max}	Computation time [s]
DLMI (2.25)	0.05	0.2	2.5
	0.025	0.25	12.7
	0.0125	0.29	257
	0.00833	0.3	1259
Solution of (2.17) and inequality (2.24)	0.003	0.345	6

different values of T_s. The obtained estimates of q_{max}, the corresponding values of T_s, and of the computation time are shown in Table 2.1. These results have been obtained by using a PC equipped with an Intel® i7-720QM processor and 4 GB of RAM.

We have then considered the problem of finding the maximum value of q satisfying condition (2.24), where $W(\cdot, \cdot)$ is the positive semidefinite solution of (2.17). In order to do that, equation (2.17) has been first integrated, with a sample time $T_s = 0.003$ s, by using the Euler forward method, and then the maximum value of q satisfying condition (2.24) has been evaluated by means of a linear search. As a result, it has been found the estimate $q_{max} = 0.345$, with a computation time of about 6s, as it is shown in the last row of Table 2.1.

We can conclude that the condition based on the Reachability Gramian is much more efficient with respect to the solution of the DLMI when considering the IO-FTS analysis problem; however, as said, the DLMI feasibility problem is necessary in order to solve the finite-time stabilization problem discussed in Chapter 3. △

2.3 A Sufficient Condition for IO-FTS for $\mathcal{W}_\infty$ Exogenous Inputs

In this section we deal with the IO-FTS of LTV systems when the class of essentially bounded signals over the time interval Ω is considered. In doing that, we can go back to the case of a nonzero feedthrough matrix-valued function $G(\cdot)$. Hence, in this section we will refer again to the more general case of a LTV system described by (1.8).

Theorem 2.4 (Sufficient condition for IO-FTS, $\mathcal{W}_\infty$ inputs [30]) Given the time interval Ω, the class of inputs $\mathcal{W}_\infty$, a continuous, positive definite matrix-valued function $Q(\cdot)$, system (1.8) is input-output finite-time stable wrt (Ω, $\mathcal{W}_\infty$, $Q(\cdot)$)), if there exist a piecewise continuously differentiable, positive definite matrix-valued function $P(\cdot)$, and a piecewise continuous scalar function $\theta(\cdot) > 1$, such that the coupled DLMI/LMIs

$$\begin{pmatrix} \dot{P}(t) + A^T(t)P(t) + P(t)A(t) & P(t)F(t) \\ F^T(t)P(t) & -R(t) \end{pmatrix} < 0 \tag{2.37a}$$

$$\theta(t)R(t) - R(t) > 2\,\theta(t)G^T(t)Q(t)G(t), \tag{2.37b}$$

$$P(t) > 2\,\theta(t)C^T(t)\widetilde{Q}(t)C(t), \tag{2.37c}$$

are satisfied over Ω, being $\widetilde{Q}(t) = (t - t_0)Q(t)$. ▲

Proof: Given $t \in \Omega$, we have

$$y(t)^T Q(t)y(t) = x^T(t)C^T(t)Q(t)C(t)x(t) + w^T(t)G^T(t)Q(t)G(t)w(t)$$
$$+ x^T(t)C^T(t)Q(t)G(t)w(t) + w^T(t)G^T(t)Q(t)C(t)x(t). \qquad (2.38)$$

Now let

$$v(t) = Q(t)^{-\frac{1}{2}}C(t)x(t) - Q(t)^{-\frac{1}{2}}G(t)w(t),$$

then (in the following the time argument is omitted for brevity)

$$v^T v = x^T C^T QCx + w^T G^T QGw - x^T C^T QGw - w^T G^T QCx,$$

which can be rewritten as

$$x^T C^T QGw + w^T G^T QCx = x^T C^T QCx + w^T G^T QGw - v^T v. \qquad (2.39)$$

Replacing (2.39) in (2.38) yields

$$y^T Qy = 2x^T C^T QCx + 2w^T G^T QGw - v^T v$$
$$\leq 2(x^T C^T QCx + w^T G^T QGw). \qquad (2.40)$$

Condition (2.40) together with (2.37b) and (2.37c) imply that

$$y^T Qy < \frac{1}{\theta}\frac{x^T Px}{t - t_0} + \frac{\theta - 1}{\theta}w^T Rw$$
$$\leq \frac{1}{\theta}\frac{x^T Px}{t - t_0} + \frac{\theta - 1}{\theta}, \qquad (2.41)$$

where we have used the fact that $\|w\|_{\infty,R} \leq 1$,

Let us assume, for the moment, that $t > t_0$. From (2.37a), and by using the same arguments as in Theorem 2.3, condition iii)$\Rightarrow$i), it turns out that inequality (2.32) holds, that is

$$\frac{d}{dt}(x^T(t)P(t)x(t)) < w^T(t)R(t)w(t). \qquad (2.42)$$

Since $w(\cdot) \in \mathcal{W}_\infty$ implies that $\|w\|_{\infty,R} \leq 1$, inequality (2.42) in turn implies

$$\frac{d}{dt}(x^T(t)P(t)x(t)) < 1. \qquad (2.43)$$

Integrating (2.43) between t_0 and t, being $x(t_0) = 0$, yields

$$x(t)^T P(t)x(t) < t - t_0. \qquad (2.44)$$

Exploiting (2.44), from (2.41) we obtain

$$y^T(t)Q(t)y(t) < 1.$$

In order to conclude the proof, let us now discuss the case $t = t_0$. In this case, since the initial state $x(t_0)$ is zero, it is straightforward to prove that condition (2.37b), through (2.38) yields

$$y^T(t_0)Q(t_0)y(t_0) < 1.$$

Remark 2.2 Although for a sufficiently large value of T the condition involving $\widetilde{Q}(t) = (t - t_0)Q(t)$ may lead to ill-conditioned problems, it is worth to notice that using the finite-time stability approach makes sense especially when dealing with time horizons that are smaller then the settling time of the considered system. It turns out that typically T does not assume large values, compared to the system dynamics. Accordingly, the values of t and Q will be scaled by similar amounts, thus avoiding ill-conditioning. If it is needed to deal with time horizons much larger than the settling time of the system, then it is probably more opportune to abandon the finite-time control approach and rely on infinite-time methodologies. ◇

As concluding remark, let us briefly comment the case where the IO-FTS wrt $\mathcal{W}_\infty$ inputs for the strictly proper LTV system (2.1) is considered. Since for system (2.1), $G(t) = 0$ for all $t \in \Omega$, it can be easily shown that the optimization scalar function $\theta(\cdot)$ in Theorem 2.4 is not needed. Indeed the constraint (2.37b) is always fulfilled, while $P(\cdot)$ can be scaled in such a way that inequality (2.37c) becomes

$$P(t) \geq C^T(t)\widetilde{Q}(t)C(t).$$

Hence, in the case of strictly proper LTV systems, the following corollary holds.

Corollary 2.3 Given the time interval Ω, the class of inputs $\mathcal{W}_\infty$, a continuous, positive definite matrix-valued function $Q(\cdot)$, system (2.1) is IO-FTS wrt $(\Omega, \mathcal{W}_\infty, Q(\cdot))$, if the following coupled DLMI/LMI

$$\begin{pmatrix} \dot{P}(t) + A(t)^T P(t) + P(t)A(t) & P(t)F(t) \\ F(t)^T P(t) & -R(t) \end{pmatrix} < 0, \quad \forall t \in \Omega \tag{2.45a}$$

$$P(t) > C(t)^T \widetilde{Q}(t)C(t), \quad \forall t \in \Omega, \tag{2.45b}$$

admits a piecewise continuously differentiable, positive definite solution $P(\cdot)$, being $\widetilde{Q}(t) = (t - t_0)Q(t)$. ▲

The following example shows an application of Corollary 2.3 to a second-order proper LTI system.

Example 2.2 Consider the linear system

$$\dot{x}(t) = \begin{pmatrix} 0 & 1 \\ -3 & -2 \end{pmatrix} x(t) + \begin{pmatrix} 1 & 0 \\ 0 & 1 \end{pmatrix} w(t) \tag{2.46a}$$

$$y(t) = \begin{pmatrix} 1 & 0 \end{pmatrix} x(t), \tag{2.46b}$$

and let

$$R = \begin{pmatrix} 1 & 0 \\ 0 & 1 \end{pmatrix}.$$

Given the time interval $[0, 1.5]$, similarly to what has been done in Example 2.1, we now exploit Corollary 2.3 in order to compute an estimate of the maximum value q_{max} of the scalar weight q, such that (2.46) is IO finite-time stable wrt $(\Omega, \mathcal{W}_\infty, q)$.

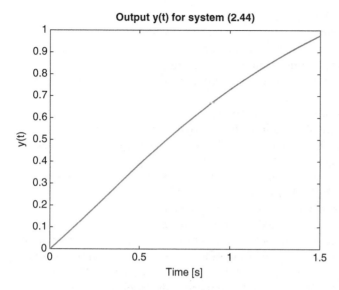

Figure 2.1 Time evolution of the output of system (2.46) when the exogenous input is set equal to $\overline{w} = \left(\frac{1}{\sqrt{2}} \quad \frac{1}{\sqrt{2}} \right)^T$.

Note that, since (2.46) has a single output, q_{max} gives an upper bound for the maximum value of $|y(t)|$ in the time interval Ω, when the exogenous signal $w(t)$ belongs to $\mathcal{W}_\infty$. In particular, such an upper bound is given by $1/\sqrt{q_{max}}$.

Once the DLMI conditions (2.45a) are turned into LMIs, by following the procedure described in Appendix C.1, with a sample time $T_s = 0.02$, an optimization problem constrained by the resulting LMIs can be solved with the MATLAB LMI Toolbox®, returning q_{max} equal to 0.727. Therefore it is possible to conclude that $|y(t)| < 1.17$ in $[0, 1.5]$. For the readers interested in implementation details, the MATLAB® script used to solve the feasibility problem (2.45a) is reported in Appendix D.1.

In order to measure the conservativeness of the sufficient condition provided by Corollary 2.3, in Figure 2.1 we show the evolution of $y(t)$ in the *worst case*, i.e., for the input $w(t) \in \mathcal{W}_\infty$, that attains the maximum value of $|y(t)|$ in Ω. In particular, taking into account that system (2.46) is LTI, it can be easily shown that such a worst case is obtained when the constant exogenous input $\overline{w} = \left(\frac{1}{\sqrt{2}} \quad \frac{1}{\sqrt{2}} \right)^T$ is considered. From Figure 2.1 it can be noticed that the maximum value of $|y(t)|$ is about 0.976. Indeed, since Corollary 2.3 gives only a sufficient condition to check IO-FTS wrt $\mathcal{W}_\infty$ signals, the value of q_{max} is underestimated. △

2.4 Summary

In this chapter we have investigated the IO-FTS problem for LTV systems. IO-FTS has been defined in the papers [30, 62]; while classical FTS [2, 4] can be considered as the finite-time counterpart of Lyapunov stability, the more recent IO-FTS concept, dealt with in this book, can be viewed as the finite-time version of the popular BIBO (or $\mathcal{L}_\infty$)-stability.

However, many substantial differences characterize the two definitions; on the one hand, BIBO stability is a structural property, since, either a system is BIBO stable or it is not, while, on the other, a given system can be both IO finite-time stable for a certain time interval, and not-IO finite-time stable for a different time interval; moreover the satisfaction of the IO-FTS property may also depend on the choice of the weighting matrices.

The IO-FTS approach is useful to refine the system behavior during the transient phase, while classical BIBO stability is a fundamental requirement to guarantee the correct behavior at steady state; therefore it is a good practice to satisfy both requirements when designing a control system. We shall discuss this issue in Chapter 10.

The main results of the chapter consist of a pair of necessary and sufficient conditions for the IO-FTS of system (1.8), when the exogenous input is assumed to belong to the family of norm bounded, square integrable signals, namely the set $\mathcal{W}_2$; in this case, for well-posedness reasons, it is assumed that the feedthrough matrix $G(\cdot)$ is zero. In particular, the former condition requires the satisfaction of a bound on the maximum eigenvalue of the positive semidefinite solution of a suitable DLE; the latter condition requires the existence of a positive definite solution to a certain coupled DLMI/LMI.

The former version of Theorem 2.3 only contained the sufficient condition for IO-FTS based on the coupled DLMI/LMI constraint, proven exploiting the properties of quadratic time-varying Lyapunov functions; such condition was published in [30]. Then the application of the Reachability Gramian theory to the IO-FTS context, allowed to demonstrate in [31] that the condition proven in [30] was also necessary, together with the equivalent condition based on the DLE.

Such conditions simplify when LTI systems are considered, since, in this case, the DLE-based constraint needs to be verified *only* at the terminal instant of the interval, while the time-varying LMI condition, coupled to the DLMI, becomes a pure terminal condition. It is important to stress that, even in the LTI case, a DLE (or a DLMI) has to be solved, in order to maintain the necessary and sufficient nature of the conditions. To this regard, when dealing with LTI systems, one could be tempted to using time-invariant Lyapunov functions; in this case, the DLMI reduces to a LMI, and the following corollary can be derived from Theorem 2.3.

Corollary 2.4 (Sufficient condition for IO-FTS of LTI systems, $\mathcal{W}_2$ inputs; time-invariant Lyapunov functions) Given the time interval Ω, the class of inputs $\mathcal{W}_2$, a positive definite matrix Q, if the coupled LMIs

$$\begin{pmatrix} A^T P + PA & PF \\ F^T P & -R \end{pmatrix} < 0 \tag{2.47a}$$

$$P > C^T Q C, \tag{2.47b}$$

admits a positive definite solution P, then system (2.33) is IO finite-time stable wrt $(\Omega, \mathcal{W}_2, Q)$. ▲

The latter approach is certainly advantageous from the computational point of view, but it introduces strong conservativeness in the IO-FTS condition; to get the point, it is sufficient to notice that condition (2.47) does not exploit the information on the finite length T of the interval Ω! In other words, the condition in Corollary 2.4 is independent on the length of the interval Ω.

Another consequence of this fact is that the condition in Corollary 2.4 also guarantees asymptotic stability of system (2.33), while the condition in Theorem 2.3 does not (this is obvious, since Theorem 2.3 is necessary for IO-FTS, which is well known to be independent from classical LS (see Chapter 1).

Coming back to Theorem 2.3, a numerical example shows that condition ii), the one based on the DLE, involves a computational burden much lower than the one required by condition iii), i.e., the DLMI-based condition. On the other hand, as it will be shown in Chapter 3, condition iii) will be the starting point for the solution of the design problem.

The other result provided in this chapter concerns the derivation of a sufficient condition for IO-FTS in presence of exogenous inputs belonging to the class of essentially bounded signals over the closed interval Ω, namely $\mathcal{W}_\infty$ (see Theorem 2.4). Again, such sufficient condition requires the solution of a feasibility problem constrained by a coupled DLMI/LMIs.

Even in this case, simplified versions of Theorem 2.4 can be derived when system (1.8) is strictly proper ($G(\cdot) = 0$) and/or the system is time-invariant. Finally, note that the set $\mathcal{W}_\infty$ contains the family of constant disturbances, which were considered in the earlier papers dealing with FTB (a concept closely related to that one of IO-FTS, see [8, 37, 38], and more recently [68]).

3

Linear Time-Varying Systems: Design of IO Finite-Time Stabilizing Controllers

In this chapter some conditions for the existence of a linear memoryless state feedback controller that IO finite-time stabilizes the closed-loop system are provided. First the case of $\mathcal{W}_2$ exogenous inputs is considered; then $\mathcal{W}_\infty$ exogenous inputs are dealt with. In both cases the conditions require the solution of a feasibility problem constrained by DLMIs coupled to LMIs.

According to the theory developed in Section 2.2, when $\mathcal{W}_2$ signals are dealt with, the additional assumption that the output does not depend on the exogenous input is needed in order to state a well-posed problem.

Then the more challenging problem of designing a dynamic output feedback controller is investigated; again necessary and sufficient conditions for the existence of a dynamical output feedback controller are derived in terms of a DLMI-based optimization problem.

More precisely, the following design problems will be considered.

Problem 3.1 (IO finite-time stabilization via State Feedback) Consider the LTV system

$$\dot{x}(t) = A(t)x(t) + B(t)u(t) + F(t)w(t), \quad x(t_0) = 0 \tag{3.1a}$$

$$y(t) = C(t)x(t) + D(t)u(t) + G(t)w(t), \tag{3.1b}$$

where $x(t) \in \mathbb{R}^n$, $u(t) \in \mathbb{R}^q$ is the control input, $w(t) \in \mathbb{R}^m$ is the exogenous input, $y(t) \in \mathbb{R}^p$ is the output, and all the involved matrix functions are piecewise continuous. Given the interval Ω, a class of disturbances $\mathcal{W}$ defined over Ω, and a continuous, positive definite matrix-valued function $Q(\cdot)$, find a state feedback control law

$$u(t) = K(t)x(t), \tag{3.2}$$

where $K(\cdot)$ is a piecewise continuous matrix-valued function of compatible dimensions, such that the closed-loop system given by the connection of system (3.1) and controller (3.2), namely

$$\dot{x}(t) = A_{cl}(t)x(t) + F(t)w(t) \tag{3.3a}$$

$$y(t) = C_{cl}(t)x(t) + G(t)w(t), \tag{3.3b}$$

with $A_{cl}(t) := (A(t) + B(t)K(t))$, $C_{cl}(t) := (C(t) + D(t)K(t))$ is IO-FTS wrt $(\Omega, \mathcal{W}, Q(\cdot))$. ◇

Finite-Time Stability: An Input-Output Approach, First Edition.
Francesco Amato, Gianmaria De Tommasi, and Alfredo Pironti.
© 2018 John Wiley & Sons Ltd. Published 2018 by John Wiley & Sons Ltd.

The next problem considers the output feedback case.

Problem 3.2 (IO finite-time stabilization via Output Feedback) Consider the LTV system

$$\dot{x}(t) = A(t)x(t) + B(t)u(t) + F(t)w(t), \quad x(t_0) = 0 \tag{3.4a}$$

$$y(t) = C(t)x(t) + G(t)w(t), \tag{3.4b}$$

where $x(t) \in \mathbb{R}^n$, $u(t) \in \mathbb{R}^q$ is the control input, $w(t) \in \mathbb{R}^m$ is the exogenous input, and $y(t) \in \mathbb{R}^p$ is the output. Given the class of signals $\mathcal{W}$, and a continuous, positive definite matrix-valued function $Q(\cdot)$ defined over Ω, find a dynamic output feedback controller in the form

$$\dot{x}_c(t) = A_K(t)x_c(t) + B_K(t)y(t) \tag{3.5a}$$

$$u(t) = C_K(t)x_c(t) + D_K(t)y(t), \tag{3.5b}$$

where $x_c(t)$ has the same dimension of $x(t)$, and all the involved matrices are piecewise continuous, such that the closed-loop system obtained by the connection of (3.4) and (3.5) is IO-FTS wrt $(\Omega, \mathcal{W}, Q(\cdot))$. In particular, the closed-loop system takes the form

$$\begin{pmatrix} \dot{x}(t) \\ \dot{x}_c(t) \end{pmatrix} = \begin{pmatrix} A + BD_K C & BC_K \\ B_K C & A_K \end{pmatrix} \begin{pmatrix} x(t) \\ x_c(t) \end{pmatrix} + \begin{pmatrix} F + BD_K G \\ B_K G \end{pmatrix} w(t)$$

$$=: A_{\mathrm{CL}} \, x_{\mathrm{CL}}(t) + F_{\mathrm{CL}} \, w(t) \tag{3.6a}$$

$$y(t) = \begin{pmatrix} C & 0 \end{pmatrix} x_{\mathrm{CL}}(t) + Gw(t) =: C_{\mathrm{CL}} \, x_{\mathrm{CL}}(t) + Gw(t), \tag{3.6b}$$

where all the considered matrices depend on time, even when not explicitly written. ◇

3.1 IO Finite-Time Stabilization via State Feedback

First we state a necessary and sufficient condition for IO finite-time stabilization via state feedback when the input class $\mathcal{W}_2$ is considered. According to the discussion made at the beginning of Section 2.2, we set $G(\cdot) = 0$ in this case. The proof of the following theorem was first given in [30, 62] for strictly proper systems; here we consider the more general case where $D(\cdot) \neq 0$.

Theorem 3.1 (IO finite-time stabilization via state feedback; $\mathcal{W}_2$ case) Given system (3.1) with $G(\cdot) = 0$, and the class of disturbances $\mathcal{W}_2$, Problem 3.1 is solvable if and only if there exist a piecewise continuously differentiable, positive definite matrix-valued function $\Pi(\cdot)$, and a piecewise continuous matrix-valued function $L(\cdot)$, such that, for all $t \in \Omega$,

$$\begin{pmatrix} \Theta(t) & F(t) \\ F(t)^T & -R \end{pmatrix} < 0 \tag{3.7a}$$

$$\begin{pmatrix} \Pi(t) & \Pi(t)C(t)^T + L^T(t)D^T(t) \\ C(t)\Pi(t) + D(t)L(t) & \Xi(t) \end{pmatrix} > 0, \tag{3.7b}$$

with

$$\Theta(t) = -\dot{\Pi}(t) + \Pi(t)A(t)^T + A(t)\Pi(t) + B(t)L(t) + L(t)^T B(t)^T,$$

and $\Xi(t) = Q^{-1}(t)$. In this case the controller gain that solves Problem 3.1, for the input class $\mathcal{W}_2$, is $K(t) = L(t)\Pi(t)^{-1}$. ▲

Proof: For all $t \in \Omega$, the conditions of Theorem 2.3 for the closed-loop system (3.3) read

$$\begin{pmatrix} \dot{P}(t) + A_{cl}^T(t)P(t) + P(t)A_{cl}(t) & P(t)F(t) \\ F^T(t)P(t) & -R(t) \end{pmatrix} < 0 \tag{3.8a}$$

$$P(t) > C_{cl}^T(t)Q(t)C_{cl}(t). \tag{3.8b}$$

Let $\Pi(t) = P^{-1}(t)$; by pre- and post-multiplying (3.8a) by $\begin{pmatrix} \Pi(t) & 0 \\ 0 & I \end{pmatrix} > 0$, and by pre- and post-multiply (3.8b) by $\Pi(t)$, we have

$$\begin{pmatrix} -\dot{\Pi}(t) + \Pi(t)A_{cl}^T(t) + A_{cl}(t)\Pi(t) & F(t) \\ F^T(t) & -R \end{pmatrix} < 0, \quad \forall t \in \Omega \tag{3.9a}$$

$$\begin{pmatrix} \Pi(t) & \Pi(t)C_{cl}^T(t) \\ C_{cl}(t)\Pi(t) & \Xi(t) \end{pmatrix} > 0, \quad \forall t \in \Omega, \tag{3.9b}$$

where (3.9b) is obtained by applying the Schur complements. The proof of the theorem then readily follows by letting $L(t) = K(t)\Pi(t)$. ◇

It is now possible to introduce the following sufficient conditions to solve Problem 3.1 when the class of $\mathcal{W}_\infty$ inputs is considered. In this case we come back to the general case $G(\cdot) \neq 0$; the proof was first provided in [30, 62] in the case $D = 0$.

Theorem 3.2 (IO finite-time stabilization via state feedback; $\mathcal{W}_\infty$ case) Given the class of disturbances $\mathcal{W}_\infty$, Problem 3.1 is solvable if there exist a piecewise continuously differentiable, positive definite matrix-valued function $\Pi(\cdot)$, a piecewise continuous matrix-valued function $L(\cdot)$, and a piecewise continuous scalar function $\theta(\cdot) > 1$, such that, for all $t \in \Omega$,

$$\begin{pmatrix} \Theta(t) & F(t) \\ F^T(t) & -R \end{pmatrix} < 0 \tag{3.10a}$$

$$\theta(t)R(t) - R(t) > 2\theta(t)G^T(t)Q(t)G(t) \tag{3.10b}$$

$$\begin{pmatrix} \Pi(t) & (t-t_0)^{1/2}(\Pi(t)C^T(t) + L^T(t)D^T(t)) \\ (t-t_0)^{1/2}(C(t)\Pi(t) + D(t)L(t)) & (2\theta(t)Q(t))^{-1} \end{pmatrix} > 0 \tag{3.10c}$$

with

$$\Theta(t) = -\dot{\Pi}(t) + \Pi(t)A^T(t) + A(t)\Pi(t) + B(t)L(t) + L^T(t)B^T(t).$$

In this case the controller gain that solves Problem 3.1, for the input class $\mathcal{W}_\infty$, is $K(t) = L(t)\Pi(t)^{-1}$. ▲

Proof: Condition (3.10a) can be obtained by following the same guidelines of Theorem 3.1.

Now, recalling that $C_{cl}(t) = C(t) + D(t)K(t)$, and by using the properties of Schur complements, we have that (3.10c) is equivalent to

$$\Pi(t) - 2\theta(t)C_{cl}^T(t)\tilde{Q}(t)C_{cl}(t) > 0, \tag{3.11}$$

where $\tilde{Q}(t) = (t - t_0)Q(t)$.

Condition (3.11) coincides with condition (2.37c), rewritten for the closed-loop system (3.3); from this the proof follows. ◇

3.2 IO-Finite-Time Stabilization via Output Feedback

In this section we exploit Theorems 2.3 and 2.4 to solve Problem 3.2. In particular, a necessary and sufficient condition for the IO finite-time stabilization of system (3.1) via dynamic output feedback is provided for the $\mathcal{W}_2$ case, and a sufficient condition for the $\mathcal{W}_\infty$ case; both conditions require the solution of a DLMI/LMI-based feasibility problem.

As usual, we start by considering the $\mathcal{W}_2$ case; to this end, according to Section 2.2, we set $G(\cdot) = 0$. The proof of this result was given in [31, 63].

Theorem 3.3 (IO finite-time stabilization via output feedback; $\mathcal{W}_2$ case) Consider system (3.4) with $G(\cdot) = 0$; Problem 3.2 is solvable *if and only if* there exist two piecewise continuously differentiable, symmetric matrix-valued functions $S(\cdot)$, $T(\cdot)$, and piecewise continuous matrix-valued functions $\hat{A}_K(\cdot)$, $\hat{B}_K(\cdot)$, $\hat{C}_K(\cdot)$ and $D_K(\cdot)$ such that the following coupled DLMI/LMI is satisfied

$$\begin{pmatrix} \Theta_{11}(t) & \Theta_{12}(t) & 0 \\ \Theta_{12}^T(t) & \Theta_{22}(t) & T(t)F(t) \\ 0 & F^T(t)T(t) & -R(t) \end{pmatrix} < 0, \quad t \in \Omega \tag{3.12a}$$

$$\begin{pmatrix} \Psi_{11}(t) & \Psi_{12}(t) & 0 \\ \Psi_{12}^T(t) & S(t) & S(t)C^T(t) \\ 0 & C(t)S(t) & Q^{-1}(t) \end{pmatrix} > 0, \quad t \in \Omega \tag{3.12b}$$

where

$$\Theta_{11}(t) = -\dot{S}(t) + A(t)S(t) + S(t)A^T(t) + B(t)\hat{C}_K(t)$$
$$\qquad + \hat{C}_K^T(t)B^T(t) + F(t)R^{-1}(t)F^T(t)$$
$$\Theta_{12}(t) = A(t) + \hat{A}_K^T(t) + B(t)D_K(t)C(t) + F(t)R^{-1}(t)F^T(t)T(t)$$
$$\Theta_{22}(t) = \dot{T}(t) + T(t)A(t) + A^T(t)T(t) + \hat{B}_K(t)C(t) + C^T(t)\hat{B}_K^T(t)$$
$$\Psi_{11}(t) = T(t) - C^T(t)Q(t)C(t)$$
$$\Psi_{12}(t) = I - C^T(t)Q(t)C(t)S(t)$$

▲

Proof: From Theorem 2.3 it readily follows that system (3.6) is IO-FTS wrt $(\mathcal{W}_2, Q(\cdot), \Omega)$ if and only if there exists a symmetric matrix-valued function $P(\cdot)$, such that

$$\dot{P}(t) + A_{\text{CL}}^T(t)P(t) + P(t)A_{\text{CL}}(t)$$
$$+ P(t)F_{\text{CL}}(t)R(t)^{-1}F_{\text{CL}}^T(t)P(t) < 0, \ t \in \Omega \tag{3.13a}$$

$$P(t) > C_{\text{CL}}^T(t)Q(t)C_{\text{CL}}(t), \quad t \in \Omega. \tag{3.13b}$$

According to [98] we define[1]

$$P(t) = \begin{pmatrix} T(t) & M(t) \\ M^T(t) & U(t) \end{pmatrix}, \ P^{-1}(t) = \begin{pmatrix} S(t) & N(t) \\ N^T(t) & \star \end{pmatrix} \tag{3.14}$$

$$\Pi_1(t) = \begin{pmatrix} S(t) & I \\ N^T(t) & 0 \end{pmatrix} \quad \Pi_2(t) = \begin{pmatrix} I & T(t) \\ 0 & M^T(t) \end{pmatrix}. \tag{3.15}$$

Note that, by definition,

$$T(t)S(t) + M(t)N^T(t) = I \tag{3.16a}$$

$$S(t)\dot{T}(t)S(t) + N(t)\dot{M}^T(t)S(t) + S(t)\dot{M}(t)N^T(t)$$
$$+ N(t)\dot{U}(t)N^T(t) = -\dot{S}(t) \tag{3.16b}$$

$$P(t)\Pi_1(t) = \Pi_2(t), \tag{3.16c}$$

where equality (3.16b) can be easily derived by noticing that

$$\dot{P}^{-1}(t) = -P^{-1}(t)\dot{P}(t)P^{-1}(t).$$

We now prove that, with the given choice of $P(t)$, conditions (3.13) are equivalent to (3.12). Indeed, by pre- and post-multiplying (3.13a)–(3.13b) by $\Pi_1^T(t)$ and $\Pi_1(t)$ respectively, and taking into account (3.16) and Lemma 5.1 in [96], the proof follows once we let

$$\begin{pmatrix} S(t) & I \\ I & T(t) \end{pmatrix} > 0 \tag{3.17a}$$

$$\hat{B}_K(t) = M(t)B_K(t) + T(t)B(t)D_K(t) \tag{3.17b}$$

$$\hat{C}_K(t) = C_K(t)N^T(t) + D_K(t)C(t)S(t) \tag{3.17c}$$

$$\hat{A}_K(t) = \dot{T}(t)S(t) + \dot{M}(t)N^T(t) + M(t)A_K(t)N^T(t)$$
$$+ T(t)B(t)C_K(t)N^T(t) + M(t)B_K(t)C(t)S(t)$$
$$+ T(t)(A(t) + B(t)D_K(t)C(t))S(t). \tag{3.17d}$$

Note that (3.17a) does not need to be explicitly imposed, since it is implied by (3.12b). $\diamond$

---

1 The symbol $\star$ denotes a "do not care block".

Remark 3.1 (Controller design) Assuming that the hypotheses of Theorem 3.3 are satisfied, in order to design the controller, the following steps have to be followed:

i) Find $S(\cdot)$, $T(\cdot)$, $\hat{A}_K(\cdot)$, $\hat{B}_K(\cdot)$, $\hat{C}_K(\cdot)$ and $D_K(\cdot)$ such that (3.12) are satisfied.
ii) Let $N(t)$ be any nonsingular matrix over Ω (e.g. $N(\cdot) = I$), and define $M(t) = [I - T(t)S(t)]N^{-T}(t)$.
iii) Obtain $A_K(\cdot)$, $B_K(\cdot)$ and $C_K(\cdot)$ by inverting (3.17). $\diamond$

Example 3.1 Let us consider the N-store building subject to an earthquake as described in Section E.1.

Figure 3.1 shows the base floor velocity and displacement time traces for the uncontrolled building with base isolation system, under the assumed earthquake excitation.

Exploiting Theorem 3.3, and assuming for $S(\cdot)$ and $T(\cdot)$ the piecewise affine structure described in Appendix C.1, it is possible to find the controller matrix-valued functions $A_k(\cdot)$, $B_k(\cdot)$, $C_k(\cdot)$, and $D_k(\cdot)$ that make system (E.1) IO-FTS with respect to the parameters given in (E.2), when $\mathcal{W}_2$ exogenous inputs are considered.

As it can be seen in Figure 3.2, the control system manages to keep very small both the velocity and the displacement of the structure. The relative control force is depicted in Figure 3.3. $\triangle$

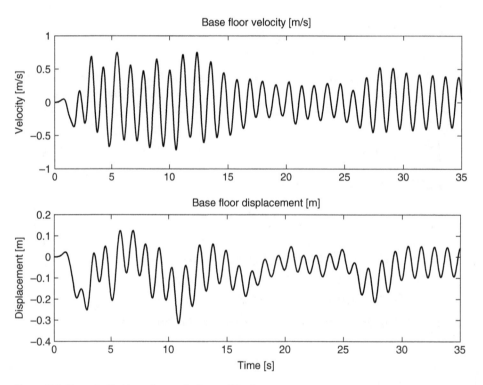

Figure 3.1 Uncontrolled base floor velocity and displacement.

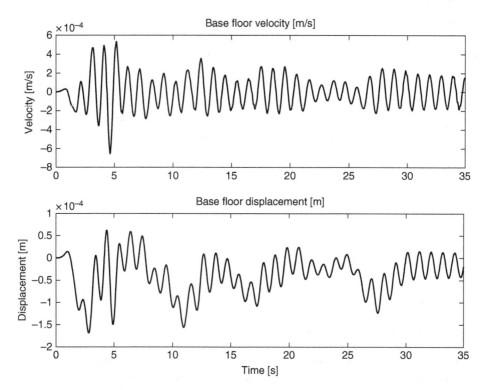

Figure 3.2 Controlled base floor velocity and displacement.

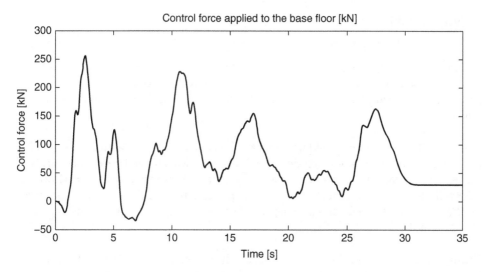

Figure 3.3 Control force applied to the base floor.

Next, we consider the design of output feedback controllers for the IO finite-time stabilization in presence of $\mathcal{W}_\infty$ inputs. The next theorem is the generalization to the case $G(\cdot) \neq 0$ of Theorem 4 in [63].

Theorem 3.4 (IO finite-time stabilization via output feedback; $\mathcal{W}_\infty$ case) Problem 3.2 is solvable *if* there exist two piecewise continuously differentiable, symmetric matrix-valued functions $S(\cdot)$, $T(\cdot)$, piecewise continuous matrix-valued functions $\hat{A}_K(\cdot)$, $\hat{B}_K(\cdot)$, $\hat{C}_K(\cdot)$ and $D_K(\cdot)$, and a piecewise continuous scalar function $\theta(\cdot) > 1$, such that the matrix inequality

$$
\begin{pmatrix}
\tilde{\Psi}_{11}(t) & \tilde{\Psi}_{12}(t) & 0 \\
\tilde{\Psi}_{12}^T(t) & S(t) & (t-t_0)^{1/2} S(t) C^T(t) \\
0 & (t-t_0)^{1/2} C(t) S(t) & (2\theta(t)Q(t))^{-1}
\end{pmatrix} > 0, \quad t \in \Omega
$$

(3.18)

where

$$
\tilde{\Psi}_{11}(t) = T(t) - 2\theta(t) C^T(t) \tilde{Q}(t) C(t),
$$
$$
\tilde{\Psi}_{12}(t) = I - 2\theta(t) C^T(t) \tilde{Q}(t) C(t) S(t),
$$
$$
\tilde{Q}(t) = (t-t_0) Q(t),
$$

is satisfied together with (3.12a) and (3.10b), for all $t \in \Omega$. ▲

Proof: From Theorem 2.4 it readily follows that system (3.6) is IO-FTS wrt $(\mathcal{W}_\infty, Q(\cdot), \Omega)$, if there exists a positive definite matrix-valued function $P(\cdot)$, and a scalar function $\theta(\cdot) > 1$, such that, for all $t \in \Omega$,

$$
\dot{P}(t) + A_{\text{CL}}^T(t)P(t) + P(t)A_{\text{CL}}(t) + P(t)F_{\text{CL}}(t)R^{-1}(t)F_{\text{CL}}^T(t)P(t) < 0,
$$

(3.19)

subject to (3.10b) and

$$
P(t) > 2\,\theta(t) C_{\text{CL}}^T(t) \tilde{Q}(t) C_{\text{CL}}(t).
$$

(3.20)

Now define $P(\cdot)$, $\Pi_1(\cdot)$, and $\Pi_2(\cdot)$ as in the proof of Theorem 3.3; by following similar arguments, we have that (3.19) is equivalent to (3.12a). Moreover, by pre- and post-multiplying (3.20) by $\Pi_1^T(t)$ and $\Pi_1(t)$ respectively, condition (3.18) follows. ◇

Remark 3.2 The reader should note that the feasibility problem stated in Theorem 3.4 is *nonlinear*, since it does not include only DLMIs and LMIs. Indeed the element $(3, 3)$ of the matrix inequality (3.18) includes the inverse of the optimization variable θ. However, as it will be shown in the next example, it is possible to turn the sufficient condition of Theorem 3.4 into a DLMI feasibility problem by setting θ equal to a constant value greater than 1. ◇

Example 3.2 In order to show how to practically exploit the condition of Theorem 3.4, let us consider the following second-order LTI system

$$
\dot{x}(t) = \begin{pmatrix} 0 & 1 \\ -3 & -2 \end{pmatrix} x(t) + \begin{pmatrix} 1 \\ 1 \end{pmatrix} u(t) + \begin{pmatrix} 0.8 & 0 \\ 0 & 0.8 \end{pmatrix} w(t), \quad x(0) = 0
$$

(3.21a)

$$
y(t) = \begin{pmatrix} 1 & 0 \end{pmatrix} x(t) + 0.05\, u(t) + \begin{pmatrix} 0.1 & 0.1 \end{pmatrix} w(t).
$$

(3.21b)

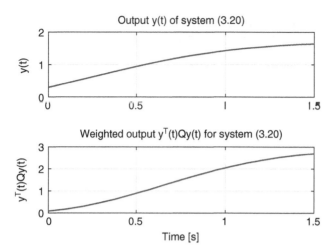

Figure 3.4 Time evolution of the output $y(\cdot)$, and the weighted output $|y(t)|_Q^2$ of system (3.21), when the exogenous input is set equal to $\overline{w} = (1.5 \ 1.5)^T$.

Given the time interval $\Omega = [0, 1.5]$ and

$$R = \begin{pmatrix} 0.2 & 0 \\ 0 & 0.2 \end{pmatrix}, \quad Q = 1,$$

system (3.21) is not IO finite-time stable wrt $(\Omega, \mathcal{W}_\infty, Q)$; as an example, Figure 3.4 shows the time traces for both the system output $y(\cdot)$ and the weighted output $|y(t)|_Q^2$ in the time interval Ω, when the exogenous input $w(\cdot)$ is set constant and equal to $\overline{w} = (1.5 \ 1.5)^T$.

In order to render system (3.21) IO finite-time stable wrt $(\Omega, \mathcal{W}_\infty, Q)$, we exploit Theorem 3.4 to design an output feedback controller that solves Problem 3.2. By setting $\overline{\theta} = 1.5$, it is possible to recast the feasibility problem of Theorem 3.4 into a DLMI feasibility problem, which can be then solved using the procedure described in Appendix C.1.

In particular, by choosing the sample time for the DLMI discretization equal to $T_s = 0.1$, the condition of Theorem 3.4 can be fulfilled. Figure 3.5 shows that, when the exogenous input $\overline{w}$ is considered, the designed output feedback is capable of keeping the weighted output within the bound. However, from Figure 3.5 it can also be observed that the control input $u(\cdot)$ assumes relatively high values at the initial time instants, and the weighted output $|y(\cdot)|_Q^2$ is kept well below 1.

It is possible to *mitigate* the control action, by adding additional LMI constraints to the feasibility problem stated in Theorem 3.4. As an example, by exploiting a similar approach as the one described in [99], the control action can be limited by adding the following LMI constraints

$$\begin{pmatrix} \lambda & \hat{C}_k^T \\ \hat{C}_k & 1 \end{pmatrix} > 0, \tag{3.22}$$

for all $t \in \Omega$, where the parameter λ permits to constrain the maximum value of the control input. Figure 3.6 shows the time traces of the weighted output $y^T(t)Qy(t)$, and of the control input $u(\cdot)$, when the constraints (3.22) are included to the feasibility problem of Theorem 3.4, with $\lambda = 100$. △

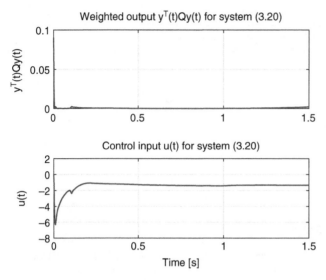

Figure 3.5 Time evolution of the weighted output $|y(t)|_Q$, and of the control input $u(\cdot)$, when the exogenous input is set equal to $\overline{w} = (1.5 \;\; 1.5)^T$, and when system (3.21) is IO finite-time stabilized by means of an output feedback controller.

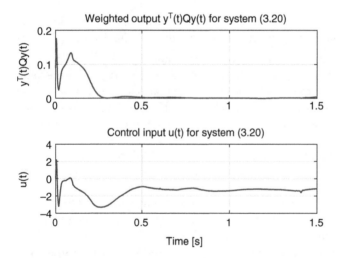

Figure 3.6 Time evolution of the weighted output $y^T(t)Qy(t)$ and of $u(\cdot)$, when the exogenous input is set equal to $\overline{w} = (1.5 \;\; 1.5)^T$, and when the output feedback controller is designed including the additional constraints (3.22) in order to limit the control input.

3.3 Summary

In this chapter, the problem of the IO finite-time stabilization has been investigated. The starting point is condition iii) in Theorem 2.3, when $\mathcal{W}_2$ exogenous inputs are considered, and Theorem 2.4 when $\mathcal{W}_\infty$ signals are dealt with.

The conditions for the existence of a state feedback controller are readily obtained through the classical change of matrix variable $K(t) = L(t)\Pi^{-1}(t)$, due to Geromel and co-workers [100], where $K(\cdot)$ is the controller gain, and $\Pi(\cdot)$ is the matrix function associated to the Lyapunov function.

When the system state is not available for feedback, the more challenging output feedback problem has been investigated. In this case, starting from condition iii) in Theorem 2.3, a necessary and sufficient condition for the IO finite-time stabilization, in presence of $\mathcal{W}_2$ exogenous inputs, has been obtained; to this end, the nonlinear change of matrix variable proposed in [98] has been exploited.

By following essentially the same machinery, a sufficient condition for stabilization, when $\mathcal{W}_\infty$ exogenous inputs are considered, has been derived starting from Theorem 3.4.

Obviously, when the involved system is time-invariant (strictly proper), simplified condition for IO finite-time stabilization, $\mathcal{W}_2$ case ($\mathcal{W}_\infty$ case), can be obtained starting from Corollary 2.2 (Corollary 2.3); further simplified conditions, at the price of strong conservativeness, can be obtained by considering time-invariant Lyapunov functions, as explained in Section 2.4.

The technique developed to face the IO finite-time stabilization problem versus $\mathcal{W}_2$ exogenous inputs has been exploited to design an active controller to smooth both displacement and speed of an N-store building subject to an earthquake.

4

IO-FTS with Nonzero Initial Conditions

In this chapter we consider a generalized definition of IO-FTS that takes into account a nonzero initial state, namely IO-FTS with nonzero initial conditions (IO-FTS-NZIC). The definition of IO-FTS-NZIC for linear systems and ellipsoidal domains has been introduced in Definition 1.5, Chapter 1.

For the sake of brevity, in this chapter we consider only the case of $\mathcal{W}_2$ exogenous inputs. We shall follow an approach similar to that one of Chapter 2, by defining the input-output operator associated to the LTV system under consideration, which also takes into account the nonzero initial condition. However, in this case, we shall see that, differently from what happens in Chapter 2, the circumstance that the weighted norm of such input-output operator is less than one, is *not* equivalent to IO-FTS-NZIC.

At the same time, we shall prove the equivalence between the condition on the norm of the operator and both a DLE-based and a DLMI-based condition. These conditions will turn to be only sufficient ones to guarantee IO-FTS-NZIC when dealing with $\mathcal{W}_2$ disturbances. As usual, the DLE-based condition is much more efficient, from the computational point of view; however, the DLMI condition will be the starting point to solve the design problem.

Indeed, the last results of the chapter will be a couple of sufficient conditions for the design of state and output feedback controllers, which render the closed-loop system IO finite-time stable NZIC.

Most of the material used in this chapter is taken by [67].

4.1   Preliminaries

In order to deal with the IO-FTS-NZIC in presence of $\mathcal{W}_2$ exogenous inputs, we consider, as usual, a strictly proper LTV system with nonzero initial condition, i.e., we deal with system (1.10) with $G(\cdot) = 0$,

$$\mathcal{LSNZ} : \begin{cases} \dot{x}(t) = A(t)x(t) + F(t)w(t), & x(t_0) = x_0 \\ y(t) = C(t)x(t). \end{cases} \tag{4.1}$$

Before presenting the main conditions to check IO-FTS-NZIC of system (4.1) with respect to $\mathcal{W}_2$ signals, some preliminary results are introduced.

Finite-Time Stability: An Input-Output Approach, First Edition.
Francesco Amato, Gianmaria De Tommasi, and Alfredo Pironti.
© 2018 John Wiley & Sons Ltd. Published 2018 by John Wiley & Sons Ltd.

First of all, we generalize the concept of Reachability Gramian given in Appendix A.4. The following lemma can be easily derived by substitution.

Lemma 4.1 Given system (4.1), a positive definite matrix Γ_0, and a continuous, positive definite matrix valued function $R(\cdot)$, the matrix-valued function $W_{nz}(\cdot, t_0)$ defined as

$$W_{nz}(t, t_0) := \Phi(t, t_0)\Gamma_0^{-1}\Phi^T(t, t_0)$$

$$+ \int_{t_0}^{t} \Phi(t, \tau)G(\tau)R^{-1}(\tau)G^T(\tau)\Phi^T(t, \tau)d\tau, \tag{4.2}$$

is the unique positive definite solution of the DLE

$$\dot{W}_{nz}(t, t_0) = A(t)W_{nz}(t, t_0) + W_{nz}(t, t_0)A^T(t) + F(t)R^{-1}(t)F^T(t), \tag{4.3a}$$

$$W_{nz}(t_0, t_0) = \Gamma_0^{-1}. \tag{4.3b}$$

▲

Similarly to what has been done in Chapter 2, in the case of LTV systems with zero initial conditions, when dealing with $\mathcal{W}_2$ signals, system (4.1) can be viewed as a linear operator $\mathcal{LSNZ}$ that maps the pair $(x_0, w(\cdot))$ into the output signal $y(\cdot)$, namely

$$\mathcal{LSNZ} : (x_0, w(\cdot)) \in \mathbb{R}^n \times \mathcal{L}_2(\Omega) \mapsto y(\cdot) \in \mathcal{L}_\infty(\Omega). \tag{4.4}$$

Let us equip the space $\mathbb{R}^n \times \mathcal{L}_2(\Omega)$ with the norm

$$\|(v, s(\cdot))\|_{2,J,R} := \left(v^T J v + \int_\Omega s^T(t)R(t)s(t)dt \right)^{1/2}$$

$$= (|v|_J^2 + \|s(\cdot)\|_{2,R}^2)^{1/2},$$

where $(v, s(\cdot)) \in \mathbb{R}^n \times \mathcal{L}_2(\Omega)$; when $J = R(\cdot) = I$, we shall write

$$\|(v, s(\cdot))\|_{2,I,I} = \|(v, s(\cdot))\|_2.$$

The norm induced on the operator $\mathcal{LSNZ}$ will be given by

$$\|\mathcal{LSNZ}\|_{\Gamma_0, R, Q} := \sup_{(x_0, w(\cdot)) \in \mathbb{R}^n \times \mathcal{L}_2(\Omega)} \frac{\|y(\cdot)\|_{\infty, Q}}{\|(x_0, w(\cdot))\|_{2, \Gamma_0, R}}$$

$$= \sup_{\|(x_0, w(\cdot))\|_{2, \Gamma_0, R} = 1} \|y(\cdot)\|_{\infty, Q}. \tag{4.5}$$

Given $z(\cdot) \in \mathcal{L}_1(\Omega)$ and $(x_0, w(\cdot)) \in \mathbb{R}^n \times \mathcal{L}_2(\Omega)$, we can consider the dual operator of $\mathcal{LSNZ}$, namely $\overline{\mathcal{LSNZ}}$, which maps signals from $\mathcal{L}_1(\Omega)$ to vector/signal pairs in $\mathbb{R}^n \times \mathcal{L}_2(\Omega)$. For such a dual operator, the following usual property holds

$$\langle z, \mathcal{LSNZ}(x_0, w(\cdot))\rangle = \langle \overline{\mathcal{LSNZ}}(z), (x_0, w(\cdot))\rangle, \tag{4.6}$$

where

$$\langle z, \mathcal{LSNZ}(x_0, w(\cdot))\rangle = \int_\Omega z^T(t) \, \mathcal{LSNZ}(x_0, w(\cdot))(t)dt$$

$$= \int_\Omega z^T(t)C(t)\Phi(t, t_0)x_0 dt + \int_\Omega z^T(t) \int_\Omega H(t, \tau)w(\tau)d\tau dt$$

$$= \left(\int_{\Omega} \Phi^T(t,t_0) C^T(t) z(t) dt \right)^T x_0$$
$$+ \int_{\Omega} \left(\int_{\Omega} H^T(t,\tau) z(t) dt \right)^T w(\tau) d\tau, \tag{4.7}$$

where $H(\cdot, \cdot)$ is the impulse response defined in (2.5).

Comparing (4.6) and (4.7), we can conclude that

$$\overline{\mathcal{LSNZ}} z = \left(\int_{\Omega} \Phi^T(\sigma,t_0) C^T(\sigma) z(\sigma) d\sigma, \ \int_{\Omega} H^T(\sigma,\tau) z(\sigma) d\sigma \right). \tag{4.8}$$

According to duality, the induced norm of $\overline{\mathcal{LSNZ}}$ is

$$\|\overline{\mathcal{LSNZ}}\|_{Q,\Gamma_0,R} = \sup_{\|z(\cdot)\|_{1,Q}=1} \|\overline{\mathcal{LSNZ}} z\|_{2,\Gamma_0,R}.$$

The next result generalizes Lemma 2.1 to the case of nonzero initial conditions.

Lemma 4.2 Given the operators $\mathcal{LSNZ}$ and $\overline{\mathcal{LSNZ}}$, the following holds true

$$\|\mathcal{LSNZ}\|_{I,I,I} =: \|\mathcal{LSNZ}\| = \|\overline{\mathcal{LSNZ}}\| := \|\overline{\mathcal{LSNZ}}\|_{I,I,I}. \tag{4.9}$$

▲

Proof: For a given $z(\cdot) \in \mathcal{L}_1(\Omega)$, $\|z(\cdot)\|_1 = 1$, we have that

$$\|\overline{\mathcal{LSNZ}} z\|_2 = \sup_{\|(x_0,v(\cdot))\|_2=1} |\langle \overline{\mathcal{LSNZ}} z, (x_0,v) \rangle|$$

$$= \sup_{\|(x_0,v(\cdot))\|_2=1} |\langle z, \mathcal{LSNZ}(x_0,v) \rangle|$$

$$= \sup_{\|(x_0,v(\cdot))\|_2=1} \left| \int_{\Omega} z^T(t) \ \mathcal{LSNZ}(x_0,v)(t) dt \right|$$

$$\leq \sup_{\|(x_0,v(\cdot))\|_2=1} \int_{\Omega} |z^T(t) \ \mathcal{LSNZ}(x_0,v)(t)| dt$$

$$\leq \|z(\cdot)\|_1 \sup_{\|(x_0,v(\cdot))\|_2=1} \|\mathcal{LSNZ}(x_0,v)(\cdot)\|_{\infty} \quad \text{in view of (1.12)}$$

$$\text{with } p = 1, \ p' = \infty$$

$$= \|z(\cdot)\|_1 \|\mathcal{LSNZ}\|. \tag{4.10}$$

Therefore, we have that $\|\overline{\mathcal{LSNZ}}\| \leq \|\mathcal{LSNZ}\|$.

Conversely, given $(x_0, w(\cdot)) \in \mathbb{R}^n \times \mathcal{L}_2(\Omega)$, $\|(x_0, w(\cdot))\|_2 = 1$, we obtain, by using similar arguments as in the proof of Lemma 2.1,

$$\|\mathcal{LSNZ}(x_0,w)\|_{\infty} = \sup_{\|y(\cdot)\|_1=1} |\langle \mathcal{LSNZ}(x_0,w), y \rangle|$$

$$= \sup_{\|y(\cdot)\|_1=1} |\langle (x_0,w), \overline{\mathcal{LSNZ}} y \rangle|$$

$$\leq \|(x_0,w(\cdot))\|_2 \sup_{\|y(\cdot)\|_1=1} \|\overline{\mathcal{LSNZ}} y\|_2. \tag{4.11}$$

Therefore we have that $\|\mathcal{LSNZ}\| \leq \|\overline{\mathcal{LSNZ}}\|$; from this the proof follows. ◇

4.2 Interpretation of the Norm of the Operator $\mathcal{LSNZ}$

The next result relates the norm of the operator $\mathcal{LSNZ}$ to the properties of the solution of the DLE (4.3).

Theorem 4.1 Given the time interval Ω, a positive definite matrix Γ_0, and two continuous, positive definite matrix-valued functions $R(\cdot)$ and $Q(\cdot)$, defined over Ω, the following statements are equivalent

i) $\|\mathcal{LSNZ}\|_{\Gamma_0,R,Q} < 1$;

ii) The positive semidefinite solution $W_{nz}(\cdot, \cdot)$ of the DLE (4.3) satisfies

$$\lambda_{\max}\left(Q^{\frac{1}{2}}(t)C(t)W_{nz}(t,t_0)C^T(t)Q^{\frac{1}{2}}(t)\right) < 1, \tag{4.12}$$

for all $t \in \Omega$.

iii) The coupled DLMI/LMIs

$$\begin{pmatrix} \dot{P}(t) + A^T(t)P(t) + P(t)A(t) & P(t)F(t) \\ F^T(t)P(t) & -R(t) \end{pmatrix} < 0 \tag{4.13a}$$

$$P(t) > C^T(t)Q(t)C(t), \tag{4.13b}$$

$$P(t_0) < \Gamma_0. \tag{4.13c}$$

admits a piecewise continuously differentiable, positive definite solution $P(\cdot)$ over Ω. ▲

Proof: We will prove the equivalence of the three statements by showing that **i)** $\Leftrightarrow$ **ii)**, and that **ii)** $\Leftrightarrow$ **iii)**.

[i) $\Leftrightarrow$ **ii)]**. Let us introduce the following change of variables

$$\xi(t) = \Gamma_0^{\frac{1}{2}}x(t) \tag{4.14a}$$

$$v(t) = R^{\frac{1}{2}}w(t) \tag{4.14b}$$

$$\upsilon(t) = Q^{\frac{1}{2}}y(t), \tag{4.14c}$$

and consider the corresponding LTV system

$$\dot{\xi}(t) = \Gamma_0^{\frac{1}{2}}A(t)\Gamma_0^{-\frac{1}{2}}\xi(t) + \Gamma_0^{\frac{1}{2}}F(t)R(t)^{-\frac{1}{2}}v(t) =: \hat{A}(t)\xi(t) + \hat{F}(t)v(t) \tag{4.15a}$$

$$\upsilon(t) = Q(t)^{\frac{1}{2}}C(t)\Gamma_0^{-\frac{1}{2}}\xi(t) =: \hat{C}(t)\xi(t), \tag{4.15b}$$

with $\xi_0 := \xi(t_0) = \Gamma_0^{\frac{1}{2}}x_0$.

Let us denote by Ξ the linear operator associated to system (4.15), namely

$$\Xi \ : \ (\xi_0, v(\cdot)) \in \mathbb{R}^n \times \mathcal{L}_2(\Omega) \to \upsilon(\cdot) \in \mathcal{L}_\infty(\Omega),$$

and by $\overline{\Xi}$ its dual (see Figure 4.1); exploiting again duality, given $\zeta \in \mathcal{L}_1(\Omega)$, we have

$$\langle \zeta, \Xi(\xi_0, v)\rangle = \langle \overline{\Xi}(\zeta), (\xi_0, v)\rangle.$$

First, we prove that

$$\|\mathcal{LSNZ}\|_{\Gamma_0,R,Q} = \|\Xi\|,$$

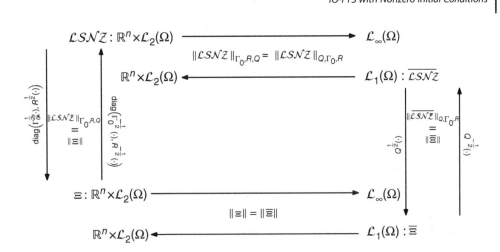

Figure 4.1 Relationships between the operators $\mathcal{LSNZ}$, Ξ and their duals.

i.e., we reduce our problem to the computation of an operator norm with unitary weighting matrices.

Indeed, we have

$$\|\mathcal{LSNZ}\|_{\Gamma_0, R, Q} = \sup_{(x_0, w(\cdot)) \in \mathbb{R}^n \times \mathcal{L}_2(\Omega)} \frac{\|y(\cdot)\|_{\infty, Q}}{\|(x_0, w(\cdot))\|_{2, \Gamma_0, R}}$$

$$= \sup_{(x_0, w(\cdot)) \in \mathbb{R}^n \times \mathcal{L}_2(\Omega)} \frac{\operatorname{ess\,sup}_{t \in \Omega}[y^T(t) Q(t) y(t)]^{\frac{1}{2}}}{(|x_0|_{\Gamma_0}^2 + \|w(\cdot)\|_{2, R}^2)^{\frac{1}{2}}}$$

$$= \sup_{(\xi_0, v(\cdot)) \in \mathbb{R}^n \times \mathcal{L}_2(\Omega)} \frac{\operatorname{ess\,sup}_{t \in \Omega}|v(t)|}{(|\xi_0|^2 + \|v(\cdot)\|_2^2)^{\frac{1}{2}}}$$

$$= \sup_{(\xi_0, v(\cdot)) \in \mathbb{R}^n \times \mathcal{L}_2(\Omega)} \frac{\|v(\cdot)\|_\infty}{\|(\xi_0, v(\cdot))\|_2} = \|\Xi\|. \tag{4.16}$$

Since, by virtue of Lemma 4.2, $\|\Xi\| = \|\overline{\Xi}\|$, the theorem is proved if we show that

$$\|\overline{\Xi}\| = \sup_{t \in \Omega} \lambda_{\max}^{\frac{1}{2}}(\hat{C}(t) \hat{W}_{nz}(t, t_0) \hat{C}^T(t)), \tag{4.17}$$

being $\lambda_{\max}(\cdot)$ the maximum eigenvalue of the argument, and $\hat{W}_{nz}(t, t_0)$ is defined as the solution of the DLE

$$\dot{\hat{W}}_{nz}(t, t_0) = \hat{A}(t) \hat{W}_{nz}(t, t_0) + \hat{W}_{nz}(t, t_0) \hat{A}(t)^T + \hat{F}(t) \hat{F}^T(t) \tag{4.18a}$$

$$\hat{W}_{nz}(t, t_0) = I. \tag{4.18b}$$

Indeed, by substitution, it is simple to verify that

$$\hat{W}_{nz}(t, t_0) = \Gamma_0^{\frac{1}{2}} W_{nz}(t, t_0) \Gamma_0^{\frac{1}{2}},$$

where $W_{nz}(t, t_0)$ is defined in (4.2); hence

$$\hat{C}(t) \hat{W}_{nz}(t, t_0) \hat{C}^T(t) = Q^{\frac{1}{2}}(t) C(t) W_{nz}(t, t_0) C^T(t) Q^{\frac{1}{2}}(t).$$

Therefore, if (4.17) holds, then we have that conditions **i)** and **ii)** of Theorem 4.1 are equivalent.

In order to prove (4.17), let us define

$$\gamma := \sup_{t \in \Omega} \lambda_{\max}^{\frac{1}{2}}(\hat{C}(t)\hat{W}_{nz}(t, t_0)\hat{C}^T(t)). \tag{4.19}$$

Having defined γ, we need to prove that

$$\|\overline{\Xi}\| = \gamma. \tag{4.20}$$

In order to do that, we proceed, *mutatis mutandis*, as in proof of Theorem 2.2. First, it is necessary to build a sequence of inputs with unit norm in $\mathcal{L}_1(\Omega)$, to be fed into the LTV system that corresponds to the dual operator $\overline{\Xi}$. Moreover, this input sequence should be such that the sequence of the norms of the corresponding output signals converges to γ.

As for Theorem 2.2, we consider the subset $\Omega' \subset \Omega$, such that, for all $t \in \Omega'$

$$\lambda_{\max}^{\frac{1}{2}}(\hat{C}(t)\hat{W}_{nz}(t, t_0)\hat{C}^T(t)) \geq \gamma - \varepsilon.$$

Given $\sigma \in \Omega'$, we denote with $h(\sigma)$ the unit norm eigenvector corresponding to the maximum eigenvalue of $\hat{C}(\sigma)\hat{W}_{nz}(\sigma, t_0)\hat{C}^T(\sigma)$, and with $u_\alpha(\cdot)$ a positive scalar function with unit norm in $\mathcal{L}_1(\Omega)$, which approaches the Dirac delta function applied in σ as $\alpha \mapsto 0$.

By applying the dual operator $\overline{\Xi}$ to the signal $h(\sigma)u_\alpha(t)$ and by taking into account (4.8), we get

$$\overline{\Xi}(h(\sigma)u_\alpha(t)) = \left(\int_\Omega \hat{\Phi}^T(t, t_0)\hat{C}^T(t)h(\sigma)u_\alpha(t)dt, \ \int_\Omega \hat{H}^T(t, \tau)h(\sigma)u_\alpha(t)dt \right),$$

where $\hat{\Phi}(\cdot, \cdot)$ and $\hat{H}(\cdot, \cdot)$ are the state transition matrix and the impulse response of system (4.15). From the property of the Dirac delta function, as $\alpha \to 0$, it follows that

$$\overline{\Xi}(h(\sigma)u_\alpha(t)) \to (\hat{\Phi}^T(\sigma, t_0)\hat{C}^T(\sigma)h(\sigma), \ \hat{H}^T(\sigma, \tau)h(\sigma)).$$

Now, by exploiting similar arguments as in the proof of Theorem 2.2 it is possible to prove that

$$\lim_{\alpha \to 0} \|\overline{\Xi}(h(\sigma)u_\alpha(t))\|_2^2 = h^T(\sigma)\hat{C}(\sigma)\hat{W}_{nz}(\sigma, t_0)\hat{C}^T(\sigma)h(\sigma),$$

where, according to Lemma 4.1, $\hat{W}_{nz}(\cdot, t_0)$ is the unique solution of the DLE (4.18).

We can conclude that

$$\lim_{\alpha \to 0} \|\overline{\Xi}(h(\sigma)u_\alpha(t))\|_2 = \lambda_{\max}^{\frac{1}{2}}(\hat{C}(\sigma)\hat{W}_{nz}(\sigma, t_0)\hat{C}^T(\sigma)) \geq \gamma - \varepsilon;$$

hence, given a sufficiently small $\eta > 0$, it is possible to choose a sufficiently small α, depending on ϵ and η, such that

$$\|\overline{\Xi}(h(\sigma)u_\alpha(t))\|_2 \geq \gamma - (\varepsilon + \eta).$$

Therefore, for all $\mu \in (0, \varepsilon + \eta)$, there exists a unit norm signal $\zeta_\mu(\cdot) \in \mathcal{L}_1(\Omega)$, such that

$$\|\Xi\| = \|\overline{\Xi}\| \geq \|\overline{\Xi} \ \zeta_\mu\|_2 \geq \gamma - \mu. \tag{4.21}$$

By letting both ε and η approaching zero, we have that $\mu \to 0$, and from (4.21)

$$\|\Xi\| \geq \gamma. \tag{4.22}$$

Finally, in order to conclude the proof, in the sequel we will show that $\|\Xi\| \leq \gamma$; let

$$\Pi(t) := \hat{W}_{nz}^{-1}(t, t_0).$$

From (4.18), we have that $\Pi(\cdot)$ satisfies

$$\dot{\Pi}(t) = -\hat{A}^T(t)\Pi(t) - \Pi(t)\hat{A}(t) - \Pi(t)\hat{F}(t)\hat{F}^T(t)\Pi(t), \tag{4.23a}$$

$$\Pi(t_0) = I. \tag{4.23b}$$

Evaluating the time derivative of $\xi(t)^T\Pi(t)\xi(t)$ along the solutions of system (4.15), we obtain

$$\frac{d}{dt}(\xi^T(t)\Pi(t)\xi(t)) = v(t)^T v(t) - \varrho^T(t)\varrho(t), \tag{4.24}$$

where $\varrho(t) = v(t) - \hat{F}^T(t)\Pi(t)\xi(t)$. Recalling (4.19), it is easy to see that

$$\gamma^2 I - \hat{C}(t)\Pi^{-1}(t)\hat{C}^T(t) \geq 0. \tag{4.25}$$

Using Schur complements, inequality (4.25) can be equivalently rewritten as

$$\begin{pmatrix} \gamma^2 I & \hat{C}(t) \\ \hat{C}^T(t) & \Pi(t) \end{pmatrix} \geq 0, \tag{4.26}$$

which, in turn, is equivalent to

$$\gamma^2\Pi(t) \geq \hat{C}^T(t)\hat{C}(t). \tag{4.27}$$

Integrating (4.24) over the time interval (t_0, t), with $t \in \Omega$, we obtain

$$\xi^T(t)\Pi(t)\xi(t) \leq \xi_0^T \xi_0 + \int_\Omega v^T(\sigma)v(\sigma)d\sigma.$$

Now, making use of (4.27), we conclude that

$$v^T(t)v(t) = \xi^T(t)\hat{C}^T(t)\hat{C}(t)\xi(t)$$
$$\leq \gamma^2\xi^T(t)\Pi(t)\xi(t)$$
$$\leq \gamma^2\left(\xi_0^T \xi_0 + \int_\Omega v^T(\sigma)v(\sigma)d\sigma\right).$$

Hence, using (4.22), we have

$$\gamma \leq \|\Xi\| \leq \gamma,$$

which concludes the proof of the equivalence between i) and ii).

[ii) $\Leftrightarrow$ iii)].

In what follows, first we use continuity arguments to show that condition ii) implies that $\|\mathcal{LSNZ}\| < 1 - \epsilon$, with ϵ sufficiently small; this allows us to relax the equality constraint in condition ii), obtaining the DLMI/LMIs condition (4.13).

Indeed, given $\epsilon > 0$, by substitution, it is possible to verify that the positive definite matrix function

$$W_{nz}^\epsilon(t, t_0) := W_{nz}(t, t_0) + \epsilon\Phi(t, t_0)\Phi^T(t, t_0) + \epsilon\int_{t_0}^t \Phi(t, \tau)\Phi^T(t, \tau)d\tau, \tag{4.28}$$

with $W_{nz}(\cdot, \cdot)$ defined in (4.2), is the solution of the matrix differential equation

$$\dot{W}_{nz}^\epsilon(t, t_0) = A(t)W_{nz}^\epsilon(t, t_0) + W_\epsilon(t, t_0)A^T(t) + F(t)R(t)^{-1}F^T(t) + \epsilon I, \tag{4.29a}$$

$$W_{nz}^\epsilon(t_0, t_0) = \Gamma_0^{-1} + \epsilon I. \tag{4.29b}$$

Therefore $W_{nz}^\epsilon(\cdot, \cdot) \to^{\epsilon \to 0} W_{nz}(\cdot, \cdot)$.

Considering condition **ii)**, and by using continuity arguments, we can conclude that $\|\mathcal{LSNZ}\| < 1$, if and only if there exists a positive definite matrix-valued function $W_{nz}^e(t, t_0)$, solution of

$$\dot{W}_{nz}^e(t, t_0) - A(t)W_{nz}^e(t, t_0) - W_{nz}^e(t, t_0)A^T(t) - F(t)R(t)^{-1}F^T(t) > 0, \tag{4.30a}$$

$$W_{nz}^e(t_0, t_0) > \Gamma_0^{-1}, \tag{4.30b}$$

such that

$$\lambda_{\max}\left(Q^{\frac{1}{2}}(t)C(t)W_{nz}^e(t, t_0)C^T(t)Q^{\frac{1}{2}}(t)\right) < 1. \tag{4.31}$$

By letting $P(t) := (W_{nz}^e(t, t_0))^{-1}$, and following the same machinery used for inequalities (4.23)–(4.25), conditions (4.30)–(4.31) can be rewritten as

$$\dot{P}(t) + A^T(t)P(t) + P(t)A(t) + P(t)F(t)R^{-1}(t)F^T(t)P(t) < 0 \tag{4.32a}$$

$$P(t_0) < \Gamma_0 \tag{4.32b}$$

$$I - Q^{\frac{1}{2}}C(t)P^{-1}(t)C^T(t)Q^{\frac{1}{2}} > 0. \tag{4.32c}$$

The proof is concluded by applying the Schur complements to inequalities (4.32). ◇

4.3 Sufficient Conditions for IO-FTS-NZIC

Before exploiting the results introduced in the previous section, let us prove the following theorem, which gives a DLMI-based sufficient condition to check IO-FTS-NZIC for strictly proper LTV systems, when $\mathcal{W}_2$ disturbances are considered.

Theorem 4.2 (Sufficient conditions for IO-FTS-NZIC via DLMIs)   Given the time interval Ω, the class of inputs $\mathcal{W}_2$, a positive definite matrix Γ_0, a continuous, positive definite matrix-valued function $Q(\cdot)$, defined over Ω, system (4.1) is input-output finite-time stable NZIC wrt $(\Omega, \mathcal{W}_2, \Gamma_0, Q(\cdot))$, if there exists a scalar $k \in (0, 1)$, such that the coupled DLMI/LMIs

$$\begin{pmatrix} \dot{P}(t) + A^T(t)P(t) + P(t)A(t) & P(t)F(t) \\ F^T(t)P(t) & -kR(t) \end{pmatrix} < 0 \tag{4.33a}$$

$$P(t) > C^T(t)Q(t)C(t) \tag{4.33b}$$

$$P(t_0) < (1 - k)\Gamma_0, \tag{4.33c}$$

admits a piecewise continuously differentiable, positive definite solution $P(\cdot)$ over Ω. ▲

Proof: Let us consider the Lyapunov function $V(t, x) = x^T P(t)x$; by using (4.33a), the derivative along the trajectories of system (4.1) yields

$$\begin{aligned} \dot{V}(t, x) &= x^T(\dot{P}(t) + A^T(t)P(t) + P(t)A(t))x \\ &\quad + x^T P(t)F(t)w + w^T F^T(t)P(t)x \\ &< -x^T P(t)F(t)(kR)^{-1}F^T(t)P(t)x + x^T P(t)F(t)w \\ &\quad + w^T F^T(t)P(t)x \\ &= kw^T R(t)w - ((kR(t))^{1/2}w - v(t))^T((kR(t))^{1/2}w - v(t)), \end{aligned} \tag{4.34}$$

where $v(t) = (kR(t))^{-1/2}F^T(t)P(t)x$.

Integrating both sides in (4.34), for $\tau \in (t_0, t)$, $t \in \Omega$, we obtain

$$x^T(t)P(t)x(t) - x_0^T P(t_0)x_0 < k \int_{t_0}^t w^T(\tau)R(\tau)w(\tau)\,d\tau$$

$$- \int_{t_0}^t ((kR(\tau))^{1/2}w(\tau) - v(\tau))^T$$

$$((kR(\tau))^{1/2}w(\tau) - v(\tau))\,d\tau < k\|w\|_{2,R}^2. \tag{4.35}$$

Finally, for all $t \in \Omega$,

$$y^T(t)Q(t)y(t) = x^T(t)C^T(t)Q(t)C(t)x(t)$$

$$< x^T(t)P(t)x(t) \quad \text{from (4.33b)}$$

$$< x_0^T P(t_0)x_0 + k\|w\|_{2,R}^2 \quad \text{from (4.35)}$$

$$< (1-k)\|x_0\|_{\Gamma_0}^2 + k\|w\|_{2,R}^2 \quad \text{from (4.33c)}.$$

Therefore, for all $w(\cdot) \in \mathcal{W}_2$, we have that $\|x_0\|_{\Gamma_0} \leq 1$ implies $\|y(\cdot)\|_{\infty,Q} < 1$. From this the proof follows. ◇

Exploiting Theorems 4.1 and 4.2, it is now possible to derive the following result, which relates IO-FTS-NZIC to the norm of the $\mathcal{LSNZ}$ operator, and hence to the solution of the DLE (4.3).

Theorem 4.3 Given the time interval Ω, the class of inputs $\mathcal{W}_2$, a positive definite matrix Γ_0, a continuous, positive definite matrix-valued function $Q(\cdot)$, defined over Ω, the following statements are equivalent for all positive scalars h and k:

i) $\|\mathcal{LSNZ}\|_{h\Gamma_0, kR, Q} < 1$;

ii) The inequality

$$\lambda_{\max}\left(Q^{\frac{1}{2}}(t)C(t)W(t,t_0)C^T(t)Q^{\frac{1}{2}}(t)\right) < 1, \tag{4.36}$$

holds for all $t \in \Omega$, where $W(\cdot, \cdot)$ is the positive definite solution of the DLE

$$\dot{W}(t,t_0) = A(t)W(t,t_0) + W(t,t_0)A^T(t) + F(t)(kR(t))^{-1}(t)F^T(t) \tag{4.37a}$$

$$W(t_0, t_0) = (h\Gamma_0)^{-1}; \tag{4.37b}$$

iii) The coupled DLMI/LMIs

$$\begin{pmatrix} \dot{P}(t) + A^T(t)P(t) + P(t)A(t) & P(t)F(t) \\ F^T(t)P(t) & -kR(t) \end{pmatrix} < 0 \tag{4.38a}$$

$$P(t) > C^T(t)Q(t)C(t) \tag{4.38b}$$

$$P(t_0) < h\Gamma_0, \tag{4.38c}$$

admits a piecewise continuously differentiable, positive definite solution $P(\cdot)$ over Ω.
Moreover, if $h + k = 1$, each one of the three conditions i)–iii) implies that

iv) system (4.1) is input-output finite-time stable NZIC wrt $(\Omega, \mathcal{W}_2, \Gamma_0, Q(\cdot))$. ▲

Proof: The proof readily follows from Theorems 4.1 and 4.2. ◇

Example 4.1 In this example we apply Theorem 4.3 to estimate the smallest ρ for which the following third-order linear system

$$\dot{x}(t) = \begin{pmatrix} 0 & 1 & 0 \\ 0 & 0 & 1 \\ t-3 & -7 & t-5 \end{pmatrix} x(t) + \begin{pmatrix} 0.1 \\ 0.1 \\ 0.1 \end{pmatrix} w(t) \tag{4.39a}$$

$$y(t) = \begin{pmatrix} 1 & 1 & 0 \\ 0 & 0 & 4 \end{pmatrix} x(t), \tag{4.39b}$$

is IO finite-time stable NZIC wrt $(\Omega, \mathcal{W}_2, \rho \cdot I, Q)$, when

$$Q = I, \quad \Omega = [0, 4],$$

and the weighting matrix for the $\mathcal{W}_2$ exogenous input is set constant and equal to

$$R = \frac{1}{2}.$$

Having chosen $\Gamma_0 = \rho \cdot I$, it is possible to apply Theorem 4.3 to estimate the largest *ball* in the state space, such that, if we choose the initial state x_0 inside such a ball, system (4.39) is IO-FTS-NZIC in the given time interval and for the chosen weights.

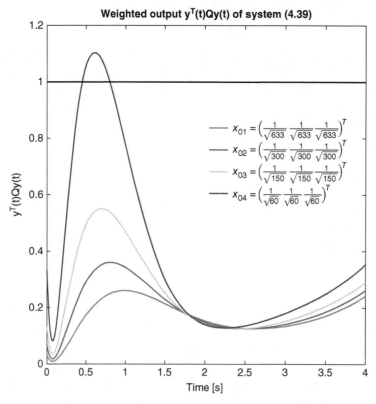

Figure 4.2 Time evolution of the weighted output $|y(\cdot)|_Q^2$ of system (4.39) for different choices of the initial state, when the exogenous input is set equal to $\overline{w} = \frac{1}{\sqrt{2}}$.

In order to estimate the smallest value for ρ, we exploit conditions **ii)** in Theorem 4.3. By setting $k \in (0, 1)$ and $h = 1 - k$, it turns out that the smallest value of ρ that allows to solve the DLE (4.37) and to fulfill inequality (4.36), is $\hat{\rho} \cong 211.23$, when $k = 0.8, h = 0.2$.

It is worth to notice that, when $h + k = 1$, condition **ii)** in Theorem 4.3 is only sufficient for IO-FTS-NZIC; in general it is $\rho_{\min} \leq \hat{\rho}$, where $\rho_{\min}$ is the smallest possible value for ρ such that system (4.39) is IO finite-time stable NZIC with respect to the chosen parameters.

In order to show this fact, Figure 4.2 reports the weighted output $|y(t)|^2_Q$ for four different values of the initial state x_0, when the exogenous input $w(t)$ is set constant and equal to $\overline{w} = \frac{1}{\sqrt{2}}$ in Ω. It can be noticed that $\|\overline{w}\|_{2,R} = 1$ and, if $\hat{\Gamma}_0 = \hat{\rho} \cdot I$, we have

$$|x_{0_1}|_{\hat{\Gamma}_0} = 1, \quad |x_{0_2}|_{\hat{\Gamma}_0} > 1, \quad |x_{0_3}|_{\hat{\Gamma}_0} > 1, \quad |x_{0_4}|_{\hat{\Gamma}_0} > 1.$$

Hence, while x_{0_4} is definitely outside the largest ball for which system (4.39) is IO-FTS-NZIC, given the simulation results and the fact that condition **iv)** is only sufficient, we cannot determine whether x_{0_2} and x_{0_3} are inside or outside such a ball. $\triangle$

4.4 Design of IO Finite-Time Stabilizing Controllers NZIC

Following the same guidelines of Chapter 3, here we shall discuss the following problems.

Problem 4.1 **(IO finite-time stabilization with NZIC via State Feedback)** Consider the LTV system

$$\dot{x}(t) = A(t)x(t) + B(t)u(t) + F(t)w(t), \quad x(t_0) = x_0 \tag{4.40a}$$
$$y(t) = C(t)x(t) + D(t)u(t), \tag{4.40b}$$

where $u(\cdot)$ is the control input, $w(\cdot)$ is the exogenous input, and all the involved matrix functions are piecewise continuous. Given the interval Ω, the class of signals $\mathcal{W}_2$, a positive definite matrix Γ_0, and a continuous, positive definite matrix-valued function $Q(\cdot)$, find a state feedback control law in the form (3.2), such that the closed-loop system given by the connection of system (4.40) and controller (3.2), namely

$$\dot{x}(t) = A_{cl}(t)x(t) + F(t)w(t), \quad x(t_0) = x_0 \tag{4.41a}$$
$$y(t) = C_{cl}(t)x(t), \tag{4.41b}$$

with $A_{cl}(t) := (A(t) + B(t)K(t))$, $C_{cl}(t) := (C(t) + D(t)K(t))$, is IO finite-time stable NZIC wrt $(\Omega, \mathcal{W}_2, \Gamma_0, Q(\cdot))$. $\triangle$

The next problem considers the output feedback case. In this circumstance, we need to introduce a further weight referred to the controller state, which we denote by Γ_K.

Problem 4.2 **(IO finite-time stabilization with NZIC via Output Feedback)** Consider the LTV system

$$\dot{x}(t) = A(t)x(t) + B(t)u(t) + F(t)w(t), \quad x(t_0) = x_0 \tag{4.42a}$$
$$y(t) = C(t)x(t), \tag{4.42b}$$

where $u(\cdot)$ is the control input, $w(\cdot)$ is the exogenous input, and all the involved matrices are piecewise continuous. Given the time interval Ω, the class of signals $\mathcal{W}_2$, two positive definite matrices Γ_0 and Γ_K, and a continuous, positive definite matrix-valued function $Q(\cdot)$ defined over Ω, find a dynamic output feedback controller in the form (3.5), such that the closed-loop system obtained by the connection of (4.42) and (3.5) is IO finite-time stable NZIC wrt $(\Omega, \mathcal{W}_2, \mathrm{diag}(\Gamma(\cdot), \Gamma_K(\cdot)), Q(\cdot))$. In particular, the closed-loop system takes the form

$$\begin{pmatrix} \dot{x}(t) \\ \dot{x}_c(t) \end{pmatrix} = \begin{pmatrix} A + BD_K C & BC_K \\ B_K C & A_K \end{pmatrix} \begin{pmatrix} x(t) \\ x_c(t) \end{pmatrix} + \begin{pmatrix} F + BD_K G \\ B_K G \end{pmatrix} w(t)$$

$$=: A_{\mathrm{CL}} \, x_{\mathrm{CL}}(t) + F_{\mathrm{CL}} \, w(t), \quad x(t_0) = x_0 \tag{4.43a}$$

$$y(t) = \begin{pmatrix} C & 0 \end{pmatrix} x_{\mathrm{CL}}(t) =: C_{\mathrm{CL}} \, x_{\mathrm{CL}}(t), \tag{4.43b}$$

where all the considered matrices depend on time, even when not explicitly written. $\triangle$

It is worth noticing that, in practical applications, the controller state is a vector of synthetic, i.e., non-physical, variables; therefore, the designer is generally not interested to bounding such variables. The introduction of the weight Γ_K is, however, needed in order to correctly state Problem 4.2.

4.4.1 State feedback

Let us consider the state feedback case first; by following the same guidelines of the proof of Theorem 3.1, and exploiting the analysis result contained in Theorem 4.2, we can prove the following result.

Theorem 4.4 (IO finite-time stabilization NZIC via state feedback) Given system (4.40), and the class of signals $\mathcal{W}_2$, Problem 4.1 is solvable if there exist a piecewise continuously differentiable, positive definite matrix-valued function $\Pi(\cdot)$, a piecewise continuous matrix-valued function $L(\cdot)$, and a scalar $k \in (0, 1)$, such that

$$\begin{pmatrix} \Theta(t) & F(t) \\ F(t)^T & -kR \end{pmatrix} < 0, \quad \forall t \in \Omega \tag{4.44a}$$

$$\begin{pmatrix} \Pi(t) & \Pi(t)C(t)^T + L^T(t)D^T(t) \\ C(t)\Pi(t) + D(t)L(t) & \Xi(t) \end{pmatrix} > 0, \quad \forall t \in \Omega$$

$$\Pi(t_0) > \frac{1}{1-k}\Gamma_0^{-1}, \tag{4.44b}$$

with

$$\Theta(t) = -\dot{\Pi}(t) + \Pi(t)A(t)^T + A(t)\Pi(t) + B(t)L(t) + L(t)^T B(t)^T,$$

and $\Xi(t) = Q(t)^{-1}$. In this case the a controller gain that solves Problem 4.1 is $K(t) = L(t)\Pi(t)^{-1}$. ▲

Example 4.2 In Example 4.1 it was shown that the linear system (4.39) is not IO finite-time stable NZIC wrt $(\Omega, \mathcal{W}_2, 20 \cdot I, I)$, when $\Omega = [0, 4]$, and the weighting matrix for the exogenous input is set equal to $R = 1/2$.

In this example, we exploit the DLMI feasibility problem stated in Theorem 4.4, in order to design a state feedback controller that renders the closed-loop system

Figure 4.3 Time evolution of the weighted output $|y(t)|_Q^2$ of the closed loop system of Example 4.2, when the initial state x_0 is taken equal to $\left(\frac{1}{\sqrt{60}} \quad \frac{1}{\sqrt{60}} \quad \frac{1}{\sqrt{60}}\right)^T$ and the exogenous input $w(t)$ is equal to $\overline{w} = \frac{1}{\sqrt{2}}$ in the interval Ω.

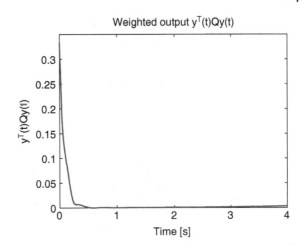

IO finite-time stable NZIC wrt $(\Omega, \mathcal{W}_2, 20 \cdot I, I)$. In particular, by setting $k = 0.1$ and $h = 0.9$, the coupled DLMI/LMIs conditions (4.44) are turned into LMIs, by using the usual piecewise affine approximation for the matrix-valued optimization variables (the sample time chosen in this case is $T_s = 0.2$ s), described in Appendix C.

Figure 4.3 shows the weighted output $|y(t)|_Q^2$ for the closed-loop system when the the initial state x_0 is equal to $\left(\frac{1}{\sqrt{60}} \quad \frac{1}{\sqrt{60}} \quad \frac{1}{\sqrt{60}}\right)^T$ and the exogenous input $w(t)$ is equal to $\overline{w} = \frac{1}{\sqrt{2}}$ in the interval Ω.   △

4.4.2 Output feedback

Now, let us consider the output feedback case; we can prove the following result.

Theorem 4.5 (IO finite-time stabilization NZIC via output feedback) Consider system (4.42); Problem 4.2 is solvable *if* there exist two piecewise continuously differentiable, symmetric matrix-valued functions $S(\cdot)$, $T(\cdot)$, and piecewise continuous matrix-valued functions $\hat{A}_K(\cdot)$, $\hat{B}_K(\cdot)$, $\hat{C}_K(\cdot)$ and $D_K(\cdot)$, such that the following inequalities are satisfied

$$\begin{pmatrix} \Theta_{11}(t) & \Theta_{12}(t) & 0 \\ \Theta_{12}^T(t) & \Theta_{22}(t) & T(t)F(t) \\ 0 & F^T(t)T(t) & -kR(t) \end{pmatrix} < 0, \quad t \in \Omega \tag{4.45a}$$

$$\begin{pmatrix} \Psi_{11}(t) & \Psi_{12}(t) & 0 \\ \Psi_{12}^T(t) & S(t) & S(t)C^T(t) \\ 0 & C(t)S(t) & Q^{-1}(t) \end{pmatrix} > 0, \quad t \in \Omega \tag{4.45b}$$

$$\begin{pmatrix} S(t_0) & I \\ I & T(t_0) \end{pmatrix} < (1-k) \begin{pmatrix} \Delta_{11} & S(t_0)\Gamma_0 \\ \Gamma_0 S(t_0) & \Gamma_0 \end{pmatrix}, \tag{4.45c}$$

where

$$\Theta_{11}(t) = -\dot{S}(t) + A(t)S(t) + S(t)A^T(t) + B(t)\hat{C}_K(t)$$
$$+ \hat{C}_K^T(t)B^T(t) + F(t)(kR(t))^{-1}(t)F^T(t)$$

$$\Theta_{12}(t) = A(t) + \hat{A}_K^T(t) + B(t)D_K(t)C(t) + F(t)(kR(t))^{-1}F^T(t)T(t)$$

$$\Theta_{22}(t) = \dot{T}(t) + T(t)A(t) + A^T(t)T(t) + \hat{B}_K(t)C(t) + C^T(t)\hat{B}_K^T(t)$$

$$\Psi_{11}(t) = T(t) - C^T(t)Q(t)C(t)$$

$$\Psi_{12}(t) = I - C^T(t)Q(t)C(t)S(t)$$

$$\Delta_{11} = S(t_0)\Gamma_0 S(t_0) + N(t_0)\Gamma_K N^T(t_0).$$

▲

Proof: By following the same guidelines of the proof of Theorem 3.3, and exploiting Theorem 4.2, we easily derive conditions (4.45a) and (4.45b). Now, let us define $P(\cdot)$, $\Pi_1(\cdot)$, and $\Pi_2(\cdot)$ according to (3.14) and (3.15) respectively.

Condition (4.33c) rewritten for the closed-loop system becomes

$$\begin{pmatrix} T(t_0) & M(t_0) \\ M^T(t_0) & U(t_0) \end{pmatrix} < (1-k) \begin{pmatrix} \Gamma_0 & 0 \\ 0 & \Gamma_K \end{pmatrix}; \tag{4.46}$$

By pre- and post-multiplying inequality (4.46) by $\Pi_1^T(t_0)$ and $\Pi_1(t_0)$ respectively, we obtain condition (4.45c). ◇

Concerning the numerical implementation of the conditions contained in the statement of Theorem 4.5, for the design of the output feedback controller, note that (4.45a) is a DLMI, condition (4.45b) is a time-varying LMI, while the initial condition (4.45c) *is not* an LMI, since it includes a quadratic term in the optimization variables $S(t_0)$ and $N(t_0)$.

To render the problem computationally tractable, it is possible either to check it *a posteriori* or to replace, at the price of some conservatism, the term Δ_{11} by

$$S(t_0)\Gamma_0^{1/2} + \Gamma_0^{1/2}S(t_0) + N(t_0)\Gamma_K^{1/2} + \Gamma_K^{1/2}N(t_0)^T - 2I,$$

which is a lower bound of Δ_{11}. This fact can be easily derived from the following inequality

$$(\Gamma_0^{1/2}S(t_0) - I)^T(\Gamma_0^{1/2}S(t_0) - I)$$
$$+ (\Gamma_K^{1/2}N(t_0)^T - I)^T(\Gamma_K^{1/2}N(t_0)^T - I) > 0.$$

In this way, condition (4.45c) is implied by

$$\begin{pmatrix} S(t_0) & I \\ I & T(t_0) \end{pmatrix}$$
$$< (1-k) \begin{pmatrix} S(t_0)\Gamma_0^{1/2} + \Gamma_0^{1/2}S(t_0) + N(t_0)\Gamma_K^{1/2} + \Gamma_K^{1/2}N(t_0)^T - 2I & S(t_0)\Gamma_0 \\ \Gamma_0 S(t_0) & \Gamma_0 \end{pmatrix}, \tag{4.47}$$

which is an LMI in the optimization variables.

4.5 Summary

In this chapter the IO-FTS problem in presence of nonzero initial conditions (IO-FTS-NZIC) has been discussed; for the sake of brevity, only the case where the exogenous inputs belong to the family of norm bounded, square integrable signals $\mathcal{W}_2$ has been considered.

This problem is strictly related to the concept of FTB discussed in the early papers [8, 37, 38], where, however, the state vector, rather than the output, is constrained to stay within a given ellipsoid, and the exogenous inputs are assumed to be constant and norm bounded. More recently, in the FTB context, the more general class of signals generated by a zero-input exo-system has been considered (see, among others, [10, 68, 101]); however, the set $\mathcal{W}_2$, considered in this chapter, is more general, since it contains the above-mentioned family of signals.

The first result of the chapter consists of the generalization of Theorem 2.2 to the case of nonzero initial conditions. In other words, we prove, in Theorem 4.1, the equivalence between the existence of a suitable solution to a DLE, or the feasibility of a certain optimization problem constrained by coupled DLMI/LMIs, and the boundedness of the norm of the operator associated to the LTV system under consideration. Such conditions, however, are not implied by the IO-FTS-NZIC of the system; in fact, they are only sufficient conditions for stability.

As usual, starting from the analysis condition based on the DLMI feasibility problem, sufficient conditions for the existence of both state and output feedback dynamical controllers are derived, all based on optimization problems constrained by coupled DLMI/LMIs.

5

IO-FTS with Constrained Control Inputs

In this chapter we introduce the concept of *structured* IO-FTS, which allows us to take into account some constraints on the amplitude of the control inputs.

Indeed, IO finite-time stabilization, as classical control advises, cannot be considered the only requirement that the closed-loop system has to satisfy; in fact, a strong performance exhibited in terms of input-output behavior, both in the finite- and the infinite-interval cases, is often accompanied by a comparable strong stress of the control variables. In practical control systems, amplitude control limitations are a necessary constraint to impose, in order to avoid energy consumption and actuators wear.

Therefore, in a real synthesis problem, the controller should be designed taking into account a limited effort of the control variables. It is important to remark that strong excursions of the control variables are typical of the transient phase; hence, the IO-FTS approach is particularly suited to deal with this specific problem.

To this regard, in this chapter, we deal with the state feedback IO-FTS problem with constrained control inputs; in order to achieve this goal, a fictitious system is built, by augmenting the output vector with the control input variables. By using such a fictitious system, the control input can be conceptually treated in the same way as the actual outputs.

However, since outputs and control inputs need to be constrained separately, Definition 1.6 is extended to that one of *structured* IO-FTS. As a by-product of this extension, we also create a framework that allows, in the general context of IO-FTS, to partition the output vector and to impose different constraints on each group of partitioned outputs.

The material discussed in this chapter is essentially taken from [55, 56].

5.1 Structured IO-FTS and Problem Statement

In this section the definition of *structured* IO-FTS and the statement of the structured IO finite-time stabilization problem via constrained state feedback are given. As we shall see, structured IO-FTS generalizes the original concept of IO-FTS; this more general definition allows us to consider additional constraints on the control variables, when solving the synthesis problem.

To motivate the idea, we shall exploit the two-degree-of-freedom quarter-car model illustrated in Appendix E.2; indeed, the IO finite-time stabilization problem, as defined in Chapter 1, does not allow to effectively deal with constraints on the control variables as in (E.7). To this end, in this section, we shall show how to modify the definition of

Finite-Time Stability: An Input-Output Approach, First Edition.
Francesco Amato, Gianmaria De Tommasi, and Alfredo Pironti.
© 2018 John Wiley & Sons Ltd. Published 2018 by John Wiley & Sons Ltd.

IO-FTS in order to take into account, during the design phase, such kind of control requirements.

Let us now introduce the concept of *structured* IO-FTS, which generalizes the original Definition 1.6. To this end, consider system (1.8); given an α-tuple of integer numbers $p_1, \ldots, p_\alpha$, where $1 < \alpha < p$, and $\sum_{i=1}^{\alpha} p_i = p$, we partition the output vector as

$$y(t) = (y_1^T(t) \cdots y_\alpha^T(t))^T, \quad t \in \Omega. \tag{5.1}$$

Note that the output partition (5.1) induces a partition of the output equation matrices

$$C(t) = (C_1^T(t) \cdots C_\alpha^T(t))^T$$
$$G(t) = (G_1^T(t) \cdots G_\alpha^T(t))^T.$$

In the original definition of IO-FTS, the output weighting is a symmetric, positive definite matrix belonging to the space $\mathbb{R}^{p \times p}$. Here we consider α continuous, positive definite weighting matrices $Q_i(t) \in \mathbb{R}^{p_i \times p_i}$, $i = 1, \ldots, \alpha$. Letting

$$Q(t) := \text{diag}(Q_1(t), \ldots, Q_\alpha(t)), \tag{5.2}$$

we introduce the following definition of *structured* IO-FTS of LTV systems.

Definition 5.1 (Structured IO-FTS)   Given the time interval Ω, a class of signals $\mathcal{W}$ defined over Ω, the output partition (5.1), and the corresponding continuous, positive definite weighting matrix $Q(\cdot)$ defined in (5.2), system (1.8) is said to be structured IO-FTS wrt $(\Omega, \mathcal{W}, Q(\cdot))$ if

$$w(\cdot) \in \mathcal{W} \Rightarrow \|y_i(\cdot)\|_{\infty, Q_i} < 1, \quad i = 1, \ldots, \alpha.$$

◇

Given Definition 5.1, it is straightforward to note that the classical Definition 1.6 can be obtained by letting $\alpha = 1$.

Note that the triplet $(\Omega, \mathcal{W}, Q(\cdot))$ contains all the information on the partition (5.1); therefore, Definition 5.1 is not ambiguous.

The first results of this chapter, namely some conditions guaranteeing that a given system is structured IO-FTS, will be given in Section 5.2. The related design problem, concerning structured IO finite-time stabilization via state feedback, will be dealt with in Section 5.3. To state precisely the last problem, consider system (3.1) and, correspondingly, given a β-tuple of integer numbers $q_1, \ldots, q_\beta$, where $\sum_{i=1}^{\beta} q_i = q$, partition the control input vector as

$$u(t) = (u_1^T(t) \cdots u_\beta^T(t))^T, \quad t \in \Omega; \tag{5.3}$$

correspondingly consider β continuous, positive definite weighting matrix-valued functions $V_i(t) \in \mathbb{R}^{q_i \times q_i}$, $i = 1, \ldots, \beta$; define

$$V(t) := \text{diag}(V_1(t), \ldots, V_\beta(t)). \tag{5.4}$$

Eventually, we consider the following partition of $D(\cdot)$, induced by (5.1),

$$D(t) = (D_1^T(t) \cdots D_\alpha^T(t))^T.$$

Problem 5.1   (Structured IO finite-time stabilization)   Consider the LTV system (3.1). Given the time interval Ω, the class of signals $\mathcal{W}$ defined over Ω, the output partition (5.1), the input partition (5.3) and the corresponding positive definite weighting matrices $Q(\cdot)$, $V(\cdot)$ defined in (5.2) and (5.4) respectively, find a state feedback control law $u(t) = K(t)x(t)$, such that the closed-loop system

$$\dot{x}(t) = (A(t) + B(t)K(t))x(t) + F(t)w(t)$$
$$=: A_{cl}x(t) + F(t)w(t), \quad x(t_0) = 0 \tag{5.5a}$$

$$\begin{pmatrix} y(t) \\ u(t) \end{pmatrix} = \begin{pmatrix} C(t) + D(t)K(t) \\ K(t) \end{pmatrix} x(t) + \begin{pmatrix} G(t) \\ 0 \end{pmatrix} w(t)$$

$$= \begin{pmatrix} C_1(t) + D_1(t)K(t) \\ \vdots \\ C_\alpha(t) + D_\alpha(t)K(t) \\ K_1(t) \\ \vdots \\ K_\beta(t) \end{pmatrix} x(t) + \begin{pmatrix} G_1(t) \\ \vdots \\ G_\alpha(t) \\ 0 \end{pmatrix} w(t), \tag{5.5b}$$

where

$$K(\cdot) = (K_1^T(\cdot) \cdots K_\beta^T(\cdot))^T, \tag{5.6}$$

is structured IO-FTS wrt $(\mathcal{W}, \text{diag}(Q(\cdot), V(\cdot)), \Omega)$. ◇

5.2 Structured IO-FTS Analysis

Let us consider the case of $\mathcal{W}_\infty$ signals first; the following is a sufficient condition for structured IO-FTS.

Theorem 5.1 (Sufficient condition for structured IO-FTS, $\mathcal{W}_\infty$ case, [56]) Given the time interval Ω, the class of signals $\mathcal{W}_\infty$, defined over Ω, the output partition (5.1), the continuous, positive definite matrix function $Q(\cdot)$ defined in (5.2), let

$$\widetilde{Q}_i(t) = (t - t_0)Q_i(t), \quad i = 1, \ldots, \alpha,$$

if there exist a piecewise continuously differentiable, positive definite matrix function $P(\cdot)$, and α piecewise continuous scalar functions $\theta_i(\cdot)$, $\theta_i(t) > 1$, $t \in \Omega$, $i = 1, \ldots, \alpha$, such that the coupled DLMI/LMIs

$$\begin{pmatrix} \dot{P}(t) + A^T(t)P(t) + P(t)A(t) & P(t)F(t) \\ F^T(t)P(t) & -R(t) \end{pmatrix} < 0 \tag{5.7a}$$

$$\theta_i(t)R(t) - R(t) > 2\,\theta_i(t)G_i^T(t)Q_i(t)G_i(t), \quad i = 1, \ldots, \alpha \tag{5.7b}$$

$$P(t) > 2\,\theta_i(t)C_i(t)^T\widetilde{Q}_i(t)C_i(t), \quad i = 1, \ldots, \alpha, \tag{5.7c}$$

are satisfied over Ω, then system (1.8) is structured IO-FTS wrt $(\Omega, \mathcal{W}_\infty, Q(\cdot))$. ▲

Proof: Given $t \in \Omega$, we have

$$y_i(t)^T Q_i(t)y_i(t) = x^T(t)C_i^T(t)Q_i(t)C_i(t)x(t) + w^T(t)G_i^T(t)Q_i(t)G_i(t)w(t)$$
$$+ x^T(t)C_i^T(t)Q_i(t)G_i(t)w(t) + w^T(t)G_i^T(t)Q_i(t)C_i(t)x(t), \tag{5.8}$$

for all $i \in \{1, \ldots, \alpha\}$. Now let

$$v_i(t) = Q_i(t)^{\frac{1}{2}}C_i(t)x(t) - Q_i(t)^{\frac{1}{2}}G_i(t)w(t),$$

then (the time argument is omitted for brevity)

$$v_i^T v_i = x^T C_i^T Q_i C_i x + w^T G_i^T Q_i G_i w - x^T C_i^T Q_i G_i w - w^T G_i^T Q_i C_i x,$$

which can be rewritten as

$$x^T C_i^T Q_i G_i w + w^T G_i^T Q_i C_i x = x^T C_i^T Q_i C_i x + w^T G_i^T Q_i G_i w - v_i^T v_i. \tag{5.9}$$

Replacing (5.9) in (5.8), we obtain

$$y_i^T Q_i y_i = 2x^T C_i^T Q_i C_i x + 2w^T G_i^T Q_i G_i w - v_i^T v_i$$
$$\leq 2(x^T C_i^T Q_i C_i x + w^T G_i^T Q_i G_i w). \tag{5.10}$$

Condition (5.10) together with (5.7b) and (5.7c) imply that

$$y_i^T Q_i y_i < \left(\frac{1}{\theta_i} \frac{x^T P x}{t - t_0} + \frac{\theta_i - 1}{\theta_i} w^T R w \right). \tag{5.11}$$

Assuming, for the moment, that $t > t_0$, we follow the same arguments of Theorem 2.4 to prove that (5.7a) implies

$$x^T(t)P(t)x(t) < t - t_0. \tag{5.12}$$

Exploiting (5.12), and recalling that $w(\cdot) \in \mathcal{W}_\infty$ implies that $\|w\|_{\infty,R} \leq 1$, from (5.11) we obtain

$$y_i^T(t)Q_i(t)y_i(t) < 1, \qquad i \in \{1, \dots, \alpha\}.$$

To conclude the proof, let us now discuss the case in which the given t coincides with the initial time t_0. In this case, since the initial state $x(t_0)$ is zero, it is straightforward to prove that condition (5.7b), together with (5.10), is sufficient to conclude that

$$y_i^T(t_0)Q_i(t_0)y_i(t_0) < 1,$$

for $i = 1, \dots, \alpha$. $\diamond$

Remark 5.1 If, for a given i, $G_i(\cdot) = 0$, it can be easily shown that the related optimization scalar function $\theta_i(\cdot)$ is not needed, since the constraint (5.7b) is always fulfilled, while $P(\cdot)$ can be scaled in such a way that inequality (5.7c) becomes

$$P(t) \geq C_i^T(t)\tilde{Q}_i(t)C_i(t). \qquad \diamond$$

Now, let us consider structured IO-FTS in presence of $\mathcal{W}_2$ signals. In this case, we have to set $G(\cdot) = 0$, in order to guarantee well-posedness of the problem, according to what was explained in Section 2.2.

Under the assumption that $G(\cdot) = 0$, the next theorem states a necessary and sufficient condition for structured IO-FTS of system (1.8), when $\mathcal{W}_2$ signals are considered.

Theorem 5.2 (Necessary and sufficient condition for structured IO-FTS, $\mathcal{W}_2$ case, [56]) Given system (1.8) with $G(\cdot) = 0$, the time interval Ω, the class of signals $\mathcal{W}_2$ defined over Ω, the output partition (5.1) and the corresponding continuous, positive definite weighting matrix function $Q(\cdot)$ defined in (5.2), system (1.8) is structured IO-FTS wrt $(\Omega, W_2, Q(\cdot))$ *if and only if* the coupled DLMI/LMI

$$\begin{pmatrix} \dot{P}(t) + A^T(t)P(t) + P(t)A(t) & P(t)F(t) \\ F^T(t)P(t) & -R(t) \end{pmatrix} < 0 \tag{5.13a}$$

$$P(t) > C_i^T(t)Q_i(t)C_i(t), \quad i = 1, \dots, \alpha, \tag{5.13b}$$

admits a piecewise continuously differentiable, positive definite solution $P(\cdot)$ over Ω.

▲

Proof: Given the output partition (5.1), system (1.8) can be considered as a collection of α fictitious systems with the same state equation (1.8a), and output equation given by

$$y_i(t) = C_i(t)x(t),$$

for $i = 1, \dots, \alpha$. The proof of the theorem readily follows by considering the result given in condition iii) of Theorem 2.3, for each one of the α fictitious systems. It is worth noticing that condition (5.13a) is not affected by the output partition, since it involves only the state equation. ◇

5.3 State Feedback Design

In this section we propose some results to solve Problem 5.1. First of all we shall deal with the case of W_∞ input signals. At the end of the section we shall let $G(\cdot) = 0$ and give a necessary and sufficient condition to solve Problem 5.1 with respect to W_2 signals. It is worth mentioning that we can always consider $D(\cdot) \neq 0$, since we assume that the control action $u(t) = K(t)x(t)$ is bounded in Ω.

Theorem 5.3 (Structured IO finite-time stabilization, W_∞ case, [56]) Given the class of signals W_∞, Problem 5.1 is solvable *if* there exist a piecewise continuously differentiable, positive definite matrix-valued function $\Pi(\cdot)$, β piecewise continuous matrix-valued functions $L_1(\cdot), \dots, L_\beta(\cdot)$, and piecewise continuous scalar functions $\lambda_i(\cdot)$, $0 < \lambda_i(t) < 1, i = 1, \dots, \alpha, t \in \Omega$, such that

$$\begin{pmatrix} \Theta(t) & F(t) \\ F^T(t) & -R(t) \end{pmatrix} < 0, \tag{5.14a}$$

$$R(t) - \lambda_i(t)R(t) > 2\, G_i^T(t)Q_i(t)G_i(t), \qquad i = 1, \dots, \alpha \tag{5.14b}$$

$$\begin{pmatrix} \Pi(t) & (t - t_0)^{1/2}(\Pi(t)C_i^T(t) + (L_1^T(t) \cdots L_\beta^T(t))D_i^T(t)) \\ (t - t_0)^{1/2}\left(C_i(t)\Pi(t) + D_i(t)\begin{pmatrix} L_1(t) \\ \vdots \\ L_\beta(t) \end{pmatrix}\right) & \dfrac{\lambda_i(t)}{2}\, Q_i^{-1}(t) \end{pmatrix} > 0$$

$$i = 1, \dots, \alpha \tag{5.14c}$$

$$\begin{pmatrix} \Pi(t) & (t - t_0)^{1/2}L_j^T(t) \\ (t - t_0)^{1/2}L_j(t) & V_j^{-1}(t) \end{pmatrix} > 0, \qquad j = 1, \dots, \beta, \tag{5.14d}$$

for all $t \in \Omega$, with

$$\Theta(t) := -\dot{\Pi}(t) + \Pi(t)A^T(t) + A(t)\Pi(t)$$
$$+ B(t)(L_1^T(t) \cdots L_\beta^T(t))^T + (L_1^T(t) \cdots L_\beta^T(t))B^T(t).$$

A controller gain that solves Problem 5.1, for the class of signals $\mathcal{W}_\infty$, is given by (5.5.6), with $K_j(t) = L_j(t)\Pi^{-1}(t)$, and $j = 1, \dots, \beta$.   ▲

Proof: Conditions (5.7) for the augmented output closed-loop system (5.5) read

$$\begin{pmatrix} \dot{P}(t) + A_{cl}^T(t)P(t) + P(t)A_{cl}(t) & P(t)F(t) \\ F^T(t)P(t) & -R(t) \end{pmatrix} < 0, \tag{5.15a}$$

$$\theta_i(t)R(t) - R(t) > 2\,\theta_i(t)G_i^T(t)Q_i(t)G_i(t), \quad i = 1, \dots, \alpha \tag{5.15b}$$

$$P(t) > 2\theta_i(t)(C_i^T(t) + K^T(t)D_i^T(t))(t - t_0)Q_i(t)(C_i(t) + D_i(t)K(t)),$$
$$i = 1, \dots, \alpha \tag{5.15c}$$

$$P(t) > K_j^T(t)(t - t_0)V_j(t)K_j(t), \quad j = 1, \dots, \beta. \tag{5.15d}$$

Note that, for the fictitious outputs $u_j(t) = K_j(t)x(t), j = 1, \dots \beta$, in (5.5b), the only constraints to be considered are (5.15d), since for these outputs there is no direct link with the vector $w(\cdot)$ (see also Remark 5.1).

Now, let us pre- and post-multiply inequality (5.15a) by $\begin{pmatrix} \Pi(t) & 0 \\ 0 & I \end{pmatrix} > 0$, where $\Pi(t) = P^{-1}(t)$. We obtain

$$\begin{pmatrix} -\dot{\Pi}(t) + \Pi(t)A_{cl}^T(t) + A_{cl}(t)\Pi(t) & F(t) \\ F^T(t) & -R(t) \end{pmatrix} < 0,$$

which turns to be equal to (5.14a), if we let $L_j(t) = K_j(t)\Pi(t), j = 1, \dots, \beta$.

Consider now condition (5.14b); it is easy to see that, if we let $\lambda_i(\cdot) = \theta_i^{-1}(\cdot)$, then (5.15b) is equivalent to (5.14b), with $0 < \lambda_i(t) < 1, t \in \Omega$.

Similarly, inequality (5.15c) is equivalent to

$$P(t) > \frac{2}{\lambda_i(t)}(C_i^T(t) + K^T(t)D_i^T(t))(t - t_0)Q_i(t)(C_i(t) + D_i(t)K(t)),$$
$$i = 1, \dots, \alpha. \tag{5.16}$$

By pre- and post-multiplying (5.16) and (5.15d) by $\Pi(t)$, we have

$$\begin{pmatrix} \Pi(t) & (t - t_0)^{1/2}(\Pi(t)C_i^T(t) + \Pi(t)K^T(t)D_i^T(t)) \\ (t - t_0)^{1/2}(C_i(t)\Pi(t) + D_i(t)K(t)\Pi(t)) & \frac{\lambda_i(t)}{2}Q_i^{-1}(t) \end{pmatrix} > 0,$$
$$i = 1, \dots, \alpha \tag{5.17a}$$

$$\begin{pmatrix} \Pi(t) & (t - t_0)^{1/2}\Pi(t)K_j^T(t) \\ (t - t_0)^{1/2}K_j(t)\Pi(t) & V_j^{-1}(t) \end{pmatrix} > 0, \quad j = 1, \dots, \beta, \tag{5.17b}$$

where (5.17a) and (5.17b) are obtained by applying the Schur complements, and are equivalent to (5.14c) and (5.14d), respectively, when letting $L_j(t) = K_j(t)\Pi(t)$, with $j = 1, \dots, \beta$.   ◇

If $G(\cdot) = 0$, exploiting similar arguments as in the previous proof, it is possible to derive the following necessary and sufficient condition to solve Problem 5.1 in the case of $\mathcal{W}_2$ signals.

Theorem 5.4 (Structured IO finite-time stabilization, $\mathcal{W}_2$ case, [56]) Given the class of signals $\mathcal{W}_2$, and $G(\cdot) = 0$, Problem 5.1 is solvable *if and only if* there exist a piecewise continuously differentiable, positive definite matrix-valued function $\Pi(\cdot)$, and β piecewise continuous matrix-valued functions $L_1(\cdot), \dots, L_\beta(\cdot)$, such that the DLMI (5.14a) and

$$\begin{pmatrix} \Pi(t) & \Pi(t)C_i^T(t) + (L_1^T(t)\cdots L_\beta^T(t))D_i^T(t) \\ C_i(t)\Pi(t) + D_i(t)(L_1^T(t)\cdots L_\beta^T(t))^T & Q_i^{-1}(t) \end{pmatrix} > 0,$$

$$i = 1, \dots, \alpha \tag{5.18a}$$

$$\begin{pmatrix} \Pi(t) & L_j^T(t) \\ L_j(t) & V_j^{-1}(t) \end{pmatrix} > 0, \qquad\qquad j = 1, \dots, \beta \tag{5.18b}$$

hold for all $t \in \Omega$.

The controller gain that solves Problem 5.1, for the class of signals $\mathcal{W}_2$, is given by (5.6) with $K_j(\cdot) = L_j(\cdot)\Pi^{-1}(\cdot)$, and $j = 1, \dots, \beta$. ▲

5.4 Design of an Active Suspension Control System Using Structured IO-FTS

In order to frame the problem of designing the active suspension control system, illustrated in Section E.2, in the context of the structured IO finite-time stabilization, according to (5.5b), let us rewrite the output equation (E.6) by explicitly including also the active force $u(\cdot)$. It follows that the *augmented* output equation becomes

$$\begin{pmatrix} y_1(t) \\ y_2(t) \\ y_3(t) \\ u(t) \end{pmatrix} = \begin{pmatrix} \dot{x}_2(t) \\ x_1(t) \\ SS \\ \dfrac{K_u x_3(t)}{g(M_s + M_u)} \\ K(t)x(t) \end{pmatrix} = \begin{pmatrix} C_1 + D_1 K(t) \\ C_2 + D_2 K(t) \\ C_3 + D_3 K(t) \\ K(t) \end{pmatrix} x(t). \tag{5.19}$$

We will design the time-varying controller $K(\cdot)$ trying to optimize the response to an *isolated bump* modeled as the $\mathcal{W}_2$ signal $w(t) = \dot{b}(t)$, where $b(t)$ describes the ground asperity

$$b(t) = \begin{cases} \dfrac{M}{2}\left(1 - \cos\left(\dfrac{2\pi V}{L}t\right)\right), & 0 \le t \le \dfrac{L}{V} \\ 0, & t > \dfrac{L}{V} \end{cases}, \tag{5.20}$$

and where $M = 0.1\ m$, and $L = 5\ m$ are the bump height and width, while $V = 45\ km/h$ is the vehicle forward velocity. The ground asperity considered for the controller design is reported in Figure 5.1.

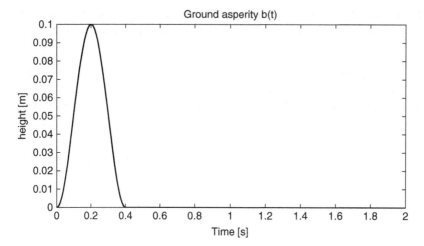

Figure 5.1 Ground asperity considered for the design of the structured IO-FTS controller for the active suspension system.

Given the bump (5.20), our goal is to minimize the body acceleration $y_1(t) = \dot{x}_2(t)$, fulfilling the constraints (E.4)–(E.7). In order to do that, we consider the following IO-FTS parameters

$$T = 2\ s, \quad R = 8.$$

Furthermore, given the selected outputs (5.19), the two outputs weighting matrices

$$Q_2 = Q_3 = 1,$$

allow to take into account the constraints (E.4) and (E.5), while the input weighting matrix is

$$V_1 = 0.15,$$

which allows to exploit the full scale of the control input when (5.20) is considered.

It turns out that, in order to minimize the body acceleration, it is possible to exploit Theorem 5.4 and solve the following optimization problem

$$\begin{aligned} &\text{minimize } \Xi_1 \\ &\text{subject to (5.18),} \end{aligned} \tag{5.21}$$

where $\Xi_1 = Q_1^{-1}$.

In order to translate the optimization problem (5.21) in terms of an LMIs optimization problem, the DLMI condition (5.18a) is recast into the LMI framework through the procedure of Appendix C.1.

Once the problem (5.21) is recast in the LMIs framework, it is possible to solve it by using off-the-shelf optimization tools such as the MATLAB LMI Toolbox® [90].

In particular, by solving (5.21), we get $\Xi_{1_{min}} = 7.22$ and the two matrix-valued functions $\Pi(\cdot)$ and $L(\cdot)$; the time-varying controller $K(\cdot)$ is then given by $K(t) = L(t)\Pi(t)^{-1}$.

It is important to remark once again that the solution of the DLMI feasibility problem is performed offline during the design phase. The interpolation between two time samples of the time-varying matrix-valued function $K(\cdot)$ is the only computation required

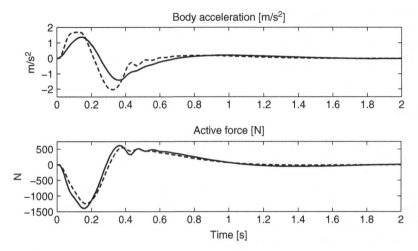

Figure 5.2 Bump response: structured IO-FTS time-varying controller (–), constrained $\mathcal{H}_\infty$ controller
(- -).

in real time to determine the control gain, which does not have any impact on the computational performance.

Figure 5.2 shows the comparison between the proposed time-varying controller $K(\cdot)$ and the constrained $\mathcal{H}_\infty$ controller proposed in [102], when the bump described by (5.20) is considered. It can be seen that, although it slightly increases the active control force, the structured IO-FTS controller reduces the maximum body acceleration for the considered bump, which is one of the parameters related to the passengers comfort. The positive acceleration is reduced from ~ 1.7 m/s² to ~ 1.35 m/s², which

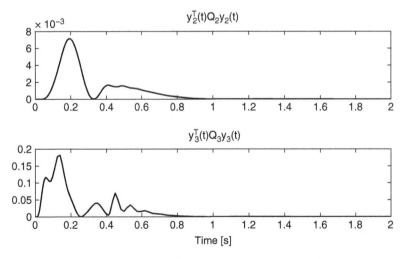

Figure 5.3 Bump response: time behavior of the weighted output $|y_2(t)|^2_{Q_2}$ and $|y_3(t)|^2_{Q_3}$ when the structured IO-FTS time-varying controller is considered.

is about 20% less. The reduction of the negative acceleration is even bigger, and is about 30% less. Furthermore, it manages to do that with an acceptable increase of control effort, which is about 15% more.

In Figure 5.3 the time behavior of $|y_2(t)|_{Q_2}^2$ and $|y_3(t)|_{Q_3}^2$ is reported.

5.5 Summary

In this Chapter, we have considered the problem of extending the IO-FTS approach described in the previous chapters with the satisfaction of some amplitude constraints on the control variables.

This methodology turns out to be useful when, as it often happens in practical design problems, there are explicit bounds on the maximum effort that can be actuated by the control system. Control limitation constraints satisfy in a quite natural way the IO-FTS requirement, since both work during the transient phase.

The proposed technique allows to deal with the control variables as if they were outputs and to specify a given constraint for each sub-vector of the (expanded) output vector; note that, as a by-product of this approach, we can constraint each one of the outputs, instead of dealing with the $y(\cdot)$ vector as a whole, differently from the classical IO-FTS definition.

A necessary and sufficient condition for structured IO finite-time stabilization is provided when $\mathcal{W}_2$ exogenous inputs are considered, while a sufficient condition is derived in presence of $\mathcal{W}_\infty$ signals. In both cases we come up to optimization problems subject to coupled DLMI/LMIs constraints.

The methodology has been exploited to design the active suspension control system, whose model is provided in Appendix E. It is worth noting that this application will be reconsidered in Chapter 10, where the design of a real control system, mixing IO-FTS, control amplitude, and Lyapunov stability requirements, will be illustrated.

6

Robustness Issues and the Mixed $\mathcal{H}_\infty$/FTS Control Problem

In this chapter we discuss the problem of the robust IO finite-time stabilization with guaranteed performance; to this regard, we consider the control scheme depicted in Figure 6.1.

In particular, given the finite-time interval Ω, our objective is to design a state feedback controller guaranteeing that, for any admissible uncertainty $\Delta(\cdot)$, and in presence of any input $w(\cdot) \in \mathcal{L}_2(\Omega)$, whose (weighted) norm is less than one, the following holds: i) the (weighted) $\mathcal{L}_2$-norm of the output $y(\cdot)$ is less than one, and ii) the peak of the (weighted) norm of the output, namely $\|y(\cdot)\|_{\infty,Q}$, is less than one (where $Q(\cdot)$ is a given positive definite matrix-function).

The former constraint is a classical robust $\mathcal{H}_\infty$ requirement involving the norm of the linear operator mapping $w(\cdot)$ to $y(\cdot)$, while the latter can be framed into the context of the IO-FTS of linear systems, dealt with in this book. Said in other words, the main contribution of this chapter consists of the investigation of the mixed $\mathcal{H}_\infty$/FTS control problem.

The rationale behind the proposed approach follows from the consideration that optimal $\mathcal{H}_\infty$ control looks at the minimization of the $\mathcal{L}_2$-norm of the output $y(\cdot)$ in Figure 6.1, that is, it tries to minimize, for all admissible exogenous input $w(\cdot)$ on the unitary sphere in $\mathcal{L}_2(\Omega)$, the *energy* of the output signal on Ω. As said, this amounts to minimize the $\mathcal{H}_\infty$-norm of the operator between $w(\cdot)$ and $y(\cdot)$.

On the other hand, energy minimization does not prevent strong output excursions, especially over finite-time intervals, where transients play a fundamental role, since the $\mathcal{H}_\infty$ approach does not account for explicit constraints on the *point-wise* value of $y(\cdot)$. Therefore the optimal $\mathcal{H}_\infty$ design is mixed with the IO-FTS control methodology which, conversely, permits to take into account requirements involving the instantaneous values of the output variable.

The $\mathcal{H}_\infty$ control theory dates back to the early Eighties, with the seminal papers by Zames, Francis, Doyle, among others, [103–106], where linear time-invariant systems were dealt with, and a frequency domain approach was exploited.

The following state-space characterization of $\mathcal{H}_\infty$ optimal controllers [107–109] and the relationships between $\mathcal{H}_\infty$ and differential games theories [110–112], allowed to extend the theory to LTV systems and deal with finite-time optimization (e. g. see [111, 113, 114]).

In this chapter we extend and mix the previous results on IO-FTS to the uncertain framework depicted in Figure 6.1. Our approach will first lead to a necessary and sufficient condition for quadratic IO-FTS (Q-IO-FTS) with an $\mathcal{H}_\infty$ bound, which in turn

Finite-Time Stability: An Input-Output Approach, First Edition.
Francesco Amato, Gianmaria De Tommasi, and Alfredo Pironti.
© 2018 John Wiley & Sons Ltd. Published 2018 by John Wiley & Sons Ltd.

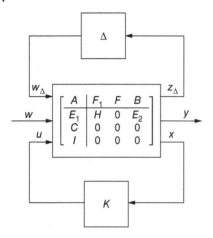

Figure 6.1 The considered state feedback control configuration.

guarantees that, for all admissible uncertainty $\Delta(\cdot)$, the open loop system in Figure 6.1 ($K = 0$): i) is IO finite-time stable, and ii) exhibits a given $\mathcal{H}_\infty$ bound between w and y. Such condition will require the solution of two DLMIs coupled with a time-varying LMI.

The necessary and sufficient condition introduced for the open loop system is then exploited to solve the synthesis problem via state-feedback. In particular, a coupled DLMIs/LMI sufficient condition is given to design a state-feedback controller that guarantees Q-IO finite-time stabilization with an $\mathcal{H}_\infty$ bound. Hence, the proposed methodology will guarantee that, for the closed loop system, the output $z(\cdot)$ is energy bounded and, at the same time, point-wise bounded.

The material presented in this chapter is essentially taken by [57].

The control configuration we refer to in this chapter is introduced at the beginning of the next section, together with the formal problem we will deal with and some preliminary results.

6.1 Preliminaries

6.1.1 System setting

Let us consider the LTV system in Figure 6.1, with $K = 0$,

$$\dot{x}(t) = A(t)x(t) + B(t)u(t) + F_1(t)w_\Delta(t) + F(t)w(t) \tag{6.1a}$$

$$z_\Delta(t) = E_1(t)x(t) + E_2(t)u(t) + H(t)w_\Delta(t) \tag{6.1b}$$

$$y(t) = C(t)x(t) \tag{6.1c}$$

$$w_\Delta(t) = \Delta(t)z_\Delta(t), \tag{6.1d}$$

where $x(t) \in \mathbb{R}^n$ is the state vector, $w_\Delta(t) \in \mathbb{R}^{m_1}$, $w(t) \in \mathbb{R}^m$, $u(t) \in \mathbb{R}^q$, $z_\Delta(t) \in \mathbb{R}^{p_1}$, $y(t) \in \mathbb{R}^p$; the matrix-valued functions in (6.1a)–(6.1c) have suitable dimensions and are assumed to be piecewise continuous.

This simplified configuration corresponds to consider only uncertainties entering the state equation. This assumption has been done to simplify the discussion; however it can be easily removed.

System (6.1) can be rewritten in compact form

$$\dot{x}(t) = (A(t) + \Delta A(t))x(t) + (B(t) + \Delta B(t))u(t) + F(t)w(t) \tag{6.2a}$$

$$y(t) = Cx(t), \tag{6.2b}$$

where

$$\big(\Delta A(t) \;\; \Delta B(t)\big) = F_1(t)\Delta(t)(I - H(t)\Delta(t))^{-1}\big(E_1(t) \;\; E_2(t)\big). \tag{6.3}$$

Remark 6.1 Note that (6.2) is well posed if $I - H(t)\Delta(t)$ is invertible for all $t \in \Omega$ with $\|\Delta(\cdot)\|_\infty \leq 1$. We will see that the conditions for Q-IO-FTS of system (6.2), provided in Section 6.2, will automatically guarantee the satisfaction of such assumption. ◇

Finally, the uncertainty block $\Delta(\cdot)$ is any piecewise continuous matrix-valued function of time, of compatible dimensions, with $\|\Delta(\cdot)\|_\infty \leq 1$; such assumption, together with non-singularity of $I - H(t)\Delta(t)$ (see Remark 6.1), guarantee existence and uniqueness of the solution of system (6.2), according to Appendix A.

6.1.2   IO-FTS with an $\mathcal{H}_\infty$ bound

In this section we introduce the mixed $\mathcal{H}_\infty$/FTS control problem with respect to the open loop *certain* system, i.e. we consider the case when $u = 0$ and $\Delta = 0$ in (6.1). A necessary and sufficient condition for the solution of such problem will be provided as a preliminary result in view of the main analysis result of the chapter. Letting $u = 0$ and $\Delta = 0$ in (6.1), we obtain

$$\dot{x}(t) = A(t)x(t) + F(t)w(t) \tag{6.4a}$$

$$y(t) = C(t)x(t). \tag{6.4b}$$

Definition 6.1 [IO-FTS with an $\mathcal{H}_\infty$ bound]   Given the time interval Ω, the set of signals $\mathcal{W}_2$, the continuous, positive definite matrix-valued functions $Q(\cdot)$, $\Gamma(\cdot)$ and $\Sigma(\cdot)$, system (6.4) is said to be IO finite-time stable wrt $(\Omega, \mathcal{W}_2, Q(\cdot))$, with an $\mathcal{H}_\infty$ bound wrt $(\Gamma(\cdot), \Sigma(\cdot))$, if, for any $w(\cdot) \in \mathcal{W}_2$,

$$\|y(\cdot)\|_{\infty,Q} < 1, \tag{6.5}$$

and

$$\sup_{\substack{w(\cdot)\in\mathcal{L}_2(\Omega) \\ \|w(\cdot)\|_{2,\Gamma}\leq 1}} \left\{ \frac{\|y(\cdot)\|_{2,\Sigma}}{\|w(\cdot)\|_{2,\Gamma}} \right\} < 1. \tag{6.6}$$

◇

The satisfaction of condition (6.5) means that system (6.4) is IO finite-time stable wrt $(\Omega, \mathcal{W}_2, Q(\cdot))$, while condition (6.6) corresponds to the fact that the (weighted) $\mathcal{H}_\infty$ norm between $w(\cdot)$ and $y(\cdot)$ is less than one.

The following lemma plays a fundamental role in the derivation of the main analysis result of this chapter. The lemma is a necessary and sufficient condition for IO-FTS with an $\mathcal{H}_\infty$ bound of system (6.4).

Lemma 6.1 (Necessary and sufficient condition for IO-FTS with an $\mathcal{H}_\infty$ bound; certain case) Given the time interval Ω, the set of signals $\mathcal{W}_2$, the continuous, positive definite matrix-valued functions $Q(\cdot)$, $\Gamma(\cdot)$ and $\Sigma(\cdot)$, system (6.4) is IO-FTS wrt $(\Omega, \mathcal{W}_2, Q(\cdot))$, with an $\mathcal{H}_\infty$ bound wrt $(\Gamma(\cdot), \Sigma(\cdot))$, *if and only if* there exist two piecewise continuously differentiable, positive definite matrix functions $P_1(\cdot)$ and $P_2(\cdot)$ that satisfy the coupled DLMIs/LMI

$$\begin{pmatrix} \dot{P}_1(t) + A^T(t)P_1(t) + P_1(t)A(t) & P_1(t)F(t) \\ F^T(t)P_1(t) & -R(t) \end{pmatrix} < 0 \tag{6.7a}$$

$$\begin{pmatrix} \dot{P}_2(t) + A^T(t)P_2(t) + P_2(t)A(t) + C^T(t)\Sigma(t)C(t) & P_2(t)F(t) \\ F^T(t)P_2(t) & -\Gamma(t) \end{pmatrix} < 0 \tag{6.7b}$$

$$P_1(t) > C^T(t)Q(t)C(t). \tag{6.7c}$$

▲

Proof: In Chapter 2, it is shown that conditions (6.7a) and (6.7c) are necessary and sufficient for the IO-FTS of system (6.4) wrt $(\Omega, \mathcal{W}_2, Q(\cdot))$. Therefore in the following we shall show that condition (6.7b) in necessary and sufficient for (6.6).

(Sufficiency) By applying Schur complements, we have that, in the time interval Ω, inequality (6.7b) is equivalent to (in the following, the time argument is omitted for the sake of brevity)

$$\dot{P}_2 + A^T P_2 + P_2 A + P_2 F\Gamma^{-1}F^T P_2 + C^T\Sigma C < 0. \tag{6.8}$$

If (6.8) holds then there exists $\varepsilon > 0$ such that

$$\dot{P}_2 + A^T P_2 + P_2 A + P_2 F\Gamma^{-1}F^T P_2 + C^T\Sigma C < -\varepsilon I,$$

and hence

$$\Xi := -(\dot{P}_2 + A^T P_2 + P_2 A + P_2 F\Gamma^{-1}F^T P_2 + C^T\Sigma C + \varepsilon I) > 0. \tag{6.9}$$

Letting

$$\bar{y} = \Xi^{1/2}x$$
$$\bar{w} = \Gamma^{1/2}w - \Gamma^{-1/2}F^T P_2 x,$$

exploiting the definition of Ξ, and taking into account that $x(t_0) = 0$, it turns out that

$$\|\bar{y}\|_2^2 = \int_\Omega x^T \Xi x \, dt$$

$$= -\int_\Omega \frac{d}{dt}(x^T P_2 x)d\tau + 2\int_\Omega w^T F^T P_2 x \; d\tau - \int_\Omega x^T P_2 F\Gamma^{-1}F^T P_2 x \; d\tau$$

$$\quad - \int_\Omega x^T C^T\Sigma C x d\tau - \varepsilon \int_\Omega x^T x d\tau$$

$$= -x^T(t_0 + T)P_2(t_0 + T)x(t_0 + T) + 2\int_\Omega w^T F^T P_2 x \; d\tau$$

$$\quad - \int_\Omega x^T P_2 F\Gamma^{-1}F^T P_2 x \; d\tau - \|y\|_{2,\Sigma}^2 - \varepsilon\|x\|_2^2; \tag{6.10}$$

moreover it can be straightforwardly checked that

$$\|w\|^2_{2,\Gamma} - \|\overline{w}\|^2_2 = -\int_\Omega x^T P_2 F \Gamma^{-1} F^T P_2 x \; d\tau$$
$$+ 2\int_\Omega w^T F^T P_2 x \; d\tau. \tag{6.11}$$

Combining (6.10) with (6.11) we obtain

$$\|y\|^2_{2,\Sigma} = \|w\|^2_{2,\Gamma} - \|\overline{w}\|^2_2 - \|\overline{y}\|^2_2 - \varepsilon\|x\|^2_2 - x^T(t_0 + T)P_2(t_0 + T)x(t_0 + T)$$
$$< \|w\|^2_{2,\Gamma}, \tag{6.12}$$

hence

$$\frac{\|y\|^2_{2,\Sigma}}{\|w\|^2_{2,\Gamma}} < 1.$$

(*Necessity*) If system (6.4) has an $\mathcal{H}_\infty$ bound wrt $(\Gamma(\cdot), \Sigma(\cdot))$, then, by continuity arguments, there exists $\delta > 0$ such that

$$\sup_{\substack{w(\cdot)\in\mathcal{L}_2(\Omega) \\ \|w(\cdot)\|_{2,\Gamma}\leq 1}} \left\{ \frac{\|y\|^2_{2,\Sigma}}{\|w\|^2_{2,\Gamma}} \right\} < 1 - \delta^2.$$

Let us now define

$$\xi := \sup_{\substack{w\in\mathcal{L}_2(\Omega) \\ \|w\|_{2,\Gamma}\leq 1}} \left\{ \frac{\|x\|_2}{\|w\|_{2,\Gamma}} \right\},$$

and

$$\tilde{y} := \begin{pmatrix} \Sigma^{1/2}C \\ \varepsilon^{1/2}I \end{pmatrix} x,$$

where $\varepsilon > 0$, such that $\varepsilon\xi^2 < \delta^2$. It follows that, for $w(\cdot)\in\mathcal{L}_2(\Omega)$, $\|w(\cdot)\|_{2,\Gamma}\leq 1$,

$$\frac{\|\tilde{y}\|^2_2}{\|w\|^2_{2,\Gamma}} = \frac{\|y\|^2_{2,\Sigma} + \varepsilon\|x\|^2_2}{\|w\|^2_{2,\Gamma}} < 1 - \delta^2 + \epsilon\xi^2 < 1,$$

which allow us to apply the equivalence between condition (*a*) and (*b*) in [114, Theorem 1.2] to the system

$$\dot{x} = Ax + F\Gamma^{-1/2}w$$
$$\tilde{y} = \begin{pmatrix} \Sigma^{1/2}C \\ \varepsilon^{1/2}I \end{pmatrix} x,$$

which implies the existence of a positive definite piecewise continuously differentiable matrix-valued function P_2, such that

$$\dot{P}_2 + A^T P_2 + P_2 A + P_2 F\Gamma^{-1}F^T P_2 + C^T\Sigma C = -\varepsilon I,$$

hence

$$\dot{P}_2 + A^T P_2 + P_2 A + P_2 F\Gamma^{-1}F^T P_2 + C^T\Sigma C < 0.$$

By applying Schur complements it is easy to show that the last inequality is equivalent to (6.7b). $\diamond$

Remark 6.2 It is interesting to notice that, in the classical Lyapunov stability context, $\mathcal{H}_\infty$ boundedness of system (6.4) also implies the IO (asymptotic) stability of the same system; indeed the satisfaction of the $\mathcal{H}_\infty$ requirement for system (6.4) is expressed by the existence of a piecewise continuously differentiable matrix function $P(\cdot)$ satisfying, fo $t \in [0, +\infty)$, the DLMI ([115], [96, Ch. 2])

$$\dot{P}(t) + A^T(t)P(t) + P(t)A(t) + P(t)F(t)F^T(t)P(t) + C^T(t)C(t) < 0.$$

The satisfaction of the above DLMI implies that $P(\cdot)$ also satisfies the DLMI

$$\dot{P}(t) + A^T(t)P(t) + P(t)A(t) < 0, \quad t \in [0, +\infty),$$

which in turn implies the IO stability of system (6.4) (see again [115], and [96, Ch. 2]). ◇

Remark 6.2 follows from the fact that, in order to satisfy the $\mathcal{H}_\infty$ requirement over an infinite interval, asymptotic stability must be necessarily guaranteed; this circumstance allows to highlight a meaningful difference between the classical Lyapunov setting and the finite-time case dealt with in this chapter.

To better investigate this point, assume that, for all $t \in \Omega$,

$$R(t) - \Gamma(t) \geq 0, \tag{6.13}$$

and that $Q = \epsilon I$, with ϵ very small with respect to the other parameters involved in the statement of Lemma 6.1.

In this case, since, from (6.7b),

$$0 > \begin{pmatrix} \dot{P}_2(t) + A^T(t)P_2(t) + P_2(t)A(t) + C^T(t)\Sigma(t)C(t) & P_2(t)F(t) \\ F^T(t)P_2(t) & -\Gamma(t) \end{pmatrix}$$

$$\geq \begin{pmatrix} \dot{P}_2(t) + A^T(t)P_2(t) + P_2(t)A(t) & P_2(t)F(t) \\ F^T(t)P_2(t) & -\Gamma(t) \end{pmatrix} \quad \text{since } C^T\Sigma C \geq 0$$

$$\geq \begin{pmatrix} \dot{P}_2(t) + A^T(t)P_2(t) + P_2(t)A(t) & P_2(t)F(t) \\ P_2(t)F^T(t) & -R(t) \end{pmatrix} \quad \text{from (6.13),} \tag{6.14}$$

it is simple to recognize that a matrix function $P_2(\cdot)$, satisfying condition (6.7b), also satisfies conditions (6.7a) and (6.7c) (due to the *smallness* of Q). In other words, in the particular case where (6.13) holds and $Q(\cdot)$ is very small, the $\mathcal{H}_\infty$ requirement *implies* the IO-FTS conditions. This is an obvious consequence of the fact that we are relaxing the point-wise constraint on the output variable.

In the general case, and differently from the Lyapunov framework recalled in Remark 6.2, condition (6.7b) does not imply (6.7a) and (6.7c), i. e. in the finite-time case considered in this chapter, $\mathcal{H}_\infty$ boundedness and IO-FTS are independent concepts (which in turn can be seen as a consequence of the fact that IO-FTS and asymptotic stability are unrelated); therefore two independent DLMIs (plus a time-varying LMI) have to be satisfied in order to guarantee both IO-FTS and $\mathcal{H}_\infty$ boundedness.

6.2 Robust and Quadratic IO-FTS with an $\mathcal{H}_\infty$ Bound

In this section we come back to the uncertain setting depicted in Figure 6.1, with $u = 0$, which leads to the LTV system (see (6.2))

$$\dot{x}(t) = (A(t) + \Delta A(t))x(t) + F(t)w(t) \tag{6.15a}$$

$$y(t) = C(t)x(t). \tag{6.15b}$$

Given the time interval Ω, the main goal of this section is to find a condition guaranteeing that:

i) System (6.15) is IO finite-time stable wrt $(\Omega, W_2, Q(\cdot))$ for all uncertainty $\Delta(\cdot)$, with $\|\Delta(\cdot)\|_\infty \leq 1$; in the following, we shall refer to such property saying that system (6.15) is *robustly* IO finite-time stable wrt (Ω, W_2, Q);

ii) The $\mathcal{H}_\infty$ norm of the operator mapping w to z in system (6.15) is less than one for all admissible realization of the uncertainty $\Delta(\cdot)$, that is (6.6) holds for all $\Delta(\cdot)$ with $\|\Delta(\cdot)\|_\infty \leq 1$; in the following, we shall refer to such property saying that system (6.15) has a *robust* $\mathcal{H}_\infty$ bound wrt $(\Gamma(\cdot), \Sigma(\cdot))$.

If conditions i) and ii) above are both satisfied, we say that *system (6.15) is robustly IO-FTS wrt $(\Omega, W_2, Q(\cdot))$ with an $\mathcal{H}_\infty$ bound wrt $(\Gamma(\cdot), \Sigma(\cdot))$.*

Since the main results of this section are derived through the use of quadratic Lyapunov functions, in the following we introduce the concept of quadratic IO-FTS with an $\mathcal{H}_\infty$ bound, which will be shown to *imply* robust IO-FTS with an $\mathcal{H}_\infty$ bound; in particular, on the basis of Lemma 6.1, the following definition and its interpretation are derived in a rather straightforward way.

Definition 6.2 Quadratic IO-FTS with an $\mathcal{H}_\infty$ bound   Given the time interval Ω, the set of signals W_2, the continuous, positive definite matrix-valued functions $Q(\cdot)$, $\Gamma(\cdot)$, and $\Sigma(\cdot)$, system (6.15) is said to be quadratically IO-FTS wrt $(\Omega, W_2, Q(\cdot))$ with an $\mathcal{H}_\infty$ bound wrt $(\Gamma(\cdot), \Sigma(\cdot))$, if there exist two piecewise continuously differentiable, positive definite matrix-valued functions $P_1(\cdot)$ and $P_2(\cdot)$ that satisfy the coupled DLMIs/LMI (time is omitted for brevity)

$$\begin{pmatrix} \dot{P}_1 + (A + \Delta A)^T P_1 + P_1(A + \Delta A) & P_1 F \\ F^T P_1 & -R \end{pmatrix} < 0 \tag{6.16a}$$

$$\begin{pmatrix} \dot{P}_2 + (A + \Delta A)^T P_2 + P_2(A + \Delta A) + C^T \Sigma C & P_2 F \\ F^T P_2 & -\Gamma \end{pmatrix} < 0, \tag{6.16b}$$

$$P_1 > C^T Q C, \tag{6.16c}$$

for all t in Ω, and for all Δ with $\|\Delta(\cdot)\|_\infty \leq 1$.   ◇

The following result, which readily follows from Lemma 6.1, gives the practical interpretation of Definition 6.2.

Lemma 6.2 If system (6.15) is quadratically IO-FTS wrt $(\Omega, W_2, Q(\cdot))$ with an $\mathcal{H}_\infty$ bound wrt $(\Gamma(\cdot), \Sigma(\cdot))$, then it is robustly IO-FTS wrt $(\Omega, W_2, Q(\cdot))$ with an $\mathcal{H}_\infty$ bound wrt $(\Gamma(\cdot), \Sigma(\cdot))$.   ▲

Remark 6.3 The term *quadratic* used in Definition 6.2 follows from the fact that Lemma 6.2 is derived by Lemma 6.1, which is proved through *quadratic* Lyapunov functions. ◇

Remark 6.4 In the certain case, see system (6.4), IO-FTS with $\mathcal{H}_\infty$ boundedness is *equivalent* to quadratic IO-FTS with $\mathcal{H}_\infty$ boundedness. In the uncertain case, dealt with in Lemma 6.2, quadratic IO-FTS with $\mathcal{H}_\infty$ boundedness *implies* robust IO-FTS with $\mathcal{H}_\infty$ boundedness, but, in general, the converse is not true. Note that, the same happens in the classical Lyapunov setting, since quadratic stability implies robust stability, but the converse is not true, as it is shown by some counterexamples (see [96, Ch. 3]). ◇

It is worth noticing that finding a solution to (6.16) is practically impossible, since (6.16a) and (6.16b) are infinite-dimensional DLMIs, that is, all possible (time-varying) realizations of $\Delta(\cdot)$ should be considered. Therefore, the main results of this chapter will consist of transforming (6.16a) and (6.16b) into a pair of finite dimensional DLMIs, *without introducing conservativeness*.

6.2.1 Main result

In this section we shall provide a computationally tractable necessary and sufficient condition for quadratic IO-FTS with an $\mathcal{H}_\infty$ bound. To achieve this goal we shall need two technical lemmas; the former gives a finite dimensional DLMI equivalent to (6.16a), while the latter gives a similar results when considering (6.16b).

Lemma 6.3 The following statements are equivalent:

i) There exists a piecewise continuously differentiable, positive definite matrix-valued function $P_1(\cdot)$ that satisfies (6.16a) for all Δ, with $\|\Delta(\cdot)\|_\infty \leq 1$.

ii) There exist a piecewise continuously differentiable, positive definite matrix-valued function $P_1(\cdot)$, and a piecewise continuous, positive definite scalar function $\mu_1(\cdot)$ which satisfy, for all $t \in \Omega$,

$$
\begin{pmatrix}
\Psi_{11}(t) & P_1(t)F(t) & \Psi_{13}(t) \\
F^T(t)P_1(t) & -R(t) & 0 \\
\Psi_{13}^T(t) & 0 & \Psi_{33}(t)
\end{pmatrix} < 0,
\tag{6.17}
$$

with

$$
\Psi_{11}(t) = \dot{P}_1(t) + A^T(t)P_1(t) + P_1(t)A(t) + \mu_1(t)E_1^T(t)E_1(t)
\tag{6.18a}
$$

$$
\Psi_{13}(t) = P_1(t)F_1(t) + \mu_1(t)E_1^T(t)H(t)
\tag{6.18b}
$$

$$
\Psi_{33}(t) = -\mu_1(t)(I - H^T(t)H(t)).
\tag{6.18c}
$$

▲

Proof: First of all, note that if (6.17) is satisfied, then the negative definiteness of $\Psi_{33}(\cdot)$ implies $I - H^T(t)H(t) > 0$, for all $t \in \Omega$. Thus we have

$$
\|H(\cdot)\Delta(\cdot)\|_\infty \leq \|H(\cdot)\|_\infty \|\Delta(\cdot)\|_\infty \leq \|H(\cdot)\|_\infty < 1,
$$

which guarantees that $I - H(t)\Delta(t)$ in (6.2) is nonsingular for all admissible $\Delta(\cdot)$.

Now, in order to prove the lemma, we show that (6.17) is equivalent to (6.16a). In particular, in the time interval Ω, inequality (6.16a) is equivalent to (the time argument is omitted for brevity)

$$\xi_1^T(\dot{P}_1 + A^T P_1 + P_1 A)\xi_1$$
$$+ \xi_1^T F_1^T [(I - H\Lambda)^T]^{-1}\Lambda^T F_1^T P_1 \xi_1$$
$$+ \xi_1^T P_1 F_1 \Lambda (I - H\Lambda)^{-1} E_1 \xi_1$$
$$+ \xi_1^T P_1 F \xi_2 + \xi_2^T F^T P_1 \xi_1 - \xi_2^T R \xi_2 < 0,$$

for all $\xi_1 \in \mathbb{R}^n, \xi_2 \in \mathbb{R}^m, \xi_1, \xi_2 \neq 0$. It follows that

$$\xi_1^T(\dot{P}_1 + A^T P_1 + P_1 A)\xi_1 - \xi_2^T R \xi_2 + \xi_1^T P_1 F \xi_2$$
$$+ \xi_2^T F^T P_1 \xi_1 + 2\xi_1^T P_1 F_1 \sigma < 0,$$

for all $\sigma \in S(\xi_1) \subseteq \mathbb{R}^{m_1}$, where

$$S(\xi_1) := \{\sigma | \sigma = \Delta(H\sigma + E_1 \xi_1), |\Delta| \leq 1\}$$
$$= \{\sigma | \sigma^T \sigma \leq (H\sigma + E_1 \xi_1)^T (H\sigma + E_1 \xi_1)\}.$$

Hence, for all t in Ω, (6.16a) is equivalent to

$$\begin{pmatrix} \xi_1 \\ \xi_2 \\ \sigma \end{pmatrix}^T \begin{pmatrix} \dot{P}_1 + A^T P_1 + P_1 A & P_1 F & P_1 F_1 \\ B_1^T F & -R & 0 \\ F_1^T P_1 & 0 & 0 \end{pmatrix} \begin{pmatrix} \xi_1 \\ \xi_2 \\ \sigma \end{pmatrix} < 0, \tag{6.19}$$

for all $\xi_1 \in \mathbb{R}^n, \xi_2 \in \mathbb{R}^m, \xi_1, \xi_2 \neq 0$, and $\sigma \in S(\xi_1)$.

Given the definition of the set $S(\xi_1)$, it follows that a positive definite, piecewise continuously differentiable matrix-function P_1 satisfies (6.16a) if and only if it satisfies (6.19) subject to the constraint

$$\begin{pmatrix} \xi_1 \\ \xi_2 \\ \sigma \end{pmatrix}^T \begin{pmatrix} -E_1^T E_1 & 0 & -E_1^T H \\ 0 & 0 & 0 \\ -H^T E_1 & 0 & I - H^T H \end{pmatrix} \begin{pmatrix} \xi_1 \\ \xi_2 \\ \sigma \end{pmatrix} \leq 0, \tag{6.20}$$

for each $\sigma \in S(\xi_1)$.

By applying S-procedure [116], this, in turn, is equivalent to the existence of a positive definite, piecewise continuously differentiable matrix-valued function $P_1(\cdot)$ and a positive scalar function $\mu_1(\cdot)$ such that (6.17) holds. $\diamond$

The following result can be derived by exploiting S-procedure arguments, as it has been done in the proof of Lemma 6.3.

Lemma 6.4 The following statements are equivalent:

i) There exists a piecewise continuously differentiable, positive definite matrix-valued function $P_2(\cdot)$ that satisfies (6.16b), for all Δ, with $\|\Delta(\cdot)\|_\infty \leq 1$.

ii) There exist a piecewise continuously differentiable, positive definite matrix-valued function $P_2(\cdot)$, and a piecewise continuous, positive definite scalar function $\mu_2(\cdot)$ which satisfy, for all $t \in \Omega$,

$$
\begin{pmatrix}
\Phi_{11}(t) & P_2(t)F(t) & \Phi_{13}(t) \\
F^T(t)P_2(t) & -\Gamma(t) & 0 \\
\Phi_{13}^T(t) & 0 & \Phi_{33}(t)
\end{pmatrix} < 0, \tag{6.21}
$$

with

$$
\Phi_{11}(t) = \dot{P}_2(t) + A^T(t)P_2(t) + P_2(t)A(t)
$$
$$
\qquad\qquad + C^T(t)\Sigma(t)C(t) + \mu_2(t)E_1^T(t)E_1(t),
$$
$$
\Phi_{13}(t) = P_2(t)F_1(t) + \mu_2(t)E_1^T(t)H(t),
$$
$$
\Phi_{33}(t) = -\mu_2(t)(I - H^T(t)H(t)).
$$

▲

Given the last two lemmas, the following result readily follows.

Theorem 6.1 (Necessary and sufficient condition for Q-IO-FTS with an $\mathcal{H}_\infty$ bound, [57]) Given the time interval Ω, three continuous, positive definite matrix-valued functions $Q(\cdot)$, $\Gamma(\cdot)$, and $\Sigma(\cdot)$, system (6.15) is Q-IO finite-time stable wrt $(\Omega, \mathcal{W}_2, Q(\cdot))$ with an $\mathcal{H}_\infty$ bound wrt $(\Gamma(\cdot), \Sigma(\cdot))$ *if and only if* there exist two piecewise continuously differentiable, positive definite matrix-valued functions $P_1(\cdot)$ and $P_2(\cdot)$, and two piecewise continuous, positive definite scalar functions $\mu_1(\cdot)$ and $\mu_2(\cdot)$, such that the coupled DLMIs/LMI (6.17), (6.16c) and (6.21) are satisfied. ▲

The considerations done at the end of Section 6.1.2 can be extended to the uncertain context. In the classical robust Lyapunov stability setting, the concept of quadratic $\mathcal{H}_\infty$ boundedness (see [117] and [96, Ch. 3]) implies quadratic stability (and hence robust stability) of the uncertain system (6.15); therefore the feasibility of a unique DLMI guarantees the satisfaction of both requirements. In the FTS context, instead, due to the independence between the $\mathcal{H}_\infty$ and the IO-FTS requirements, two decoupled DLMIs have to be satisfied (together with a time-varying LMI).

6.2.2 A numerical example

Let us consider the third order linear system

$$
A = \begin{pmatrix}
-2.0000 & -1.0000 & -0.5000 \\
2.0000 & 0 & 0 \\
0 & 1.0000 & 0
\end{pmatrix} \tag{6.22a}
$$

$$
B_1 = \begin{pmatrix} 0.5 & 0.5 & 0.5 \end{pmatrix}^T \tag{6.22b}
$$

$$
C = \begin{pmatrix}
1 & 0 & 0 \\
0 & 0 & 1
\end{pmatrix}, \tag{6.22c}
$$

and the following weighting matrices

$$R = 5, \ Q = \begin{pmatrix} 0.5 & 0 \\ 0 & 0.5 \end{pmatrix}$$
$$\Gamma = 5, \ \Sigma = \begin{pmatrix} 0.1 & 0 \\ 0 & 0.1 \end{pmatrix}. \tag{6.23}$$

When the finite-time interval $\Omega = [0, 15]$ is considered, the DLMIs/LMI feasibility problem that includes inequalities (6.17), (6.16c) and (6.21) (see the statement of Theorem 6.1) admits a solution. In particular, such feasibility problem can be casted into the LMI framework, by following the procedure described in Appendix C, and solved by using an off-the-shelf optimization software such as SeDuMi [91] or the MATLAB LMI Toolbox® [90]. To this aim, it is assumed that the optimization variables $P_1, P_2, \mu_1,$ and μ_2 in Theorem 6.1 have a piecewise affine structure.

For the system under consideration, the feasibility problem stated in Theorem 6.1 admits a solution when $T_s = 0.375$.

Let now add a norm-bounded uncertainty to system (6.22)

$$F_1 = \begin{pmatrix} 0.6 & 0.6 & 0.6 \end{pmatrix}^T$$
$$E_1 = \begin{pmatrix} 0.6 & 0.6 & 0.6 \end{pmatrix} \tag{6.24}$$
$$H = 0;$$

Figure 6.2 shows that, in this case, some realizations of the uncertainty (6.24) render the system unstable, with an high energy consumption in the time interval [0, 15] of the output variable, which means that the quadratic $\mathcal{H}_\infty$ bound cannot be guaranteed.

We shall see that the uncertain system (6.22)-(6.24) can be quadratically IO finite-time stabilized with an $\mathcal{H}_\infty$ bound, via the theory developed in the next section.

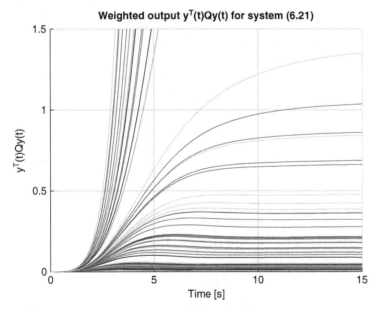

Weighted output $y^T(t)Qy(t)$ for system (6.21)

Figure 6.2 Weighted output $|y(t)|^2_Q$ for 100 random realizations of the open loop uncertain system considered in Section 6.2.2.

6.3 State Feedback Design

Here we consider the stabilization problem; given Definition 6.2, the mixed $\mathcal{H}_\infty$/FTS control problem via state feedback is defined as follows.

Problem 6.1 [**Quadratic IO finite-time stabilization**] Given the time interval Ω, and three continuous, positive definite matrix-valued functions $Q(\cdot)$, $\Gamma(\cdot)$, $\Sigma(\cdot)$, find a state feedback control law in the form (3.2), such that the closed loop system, given by the connection of system (6.2) and the controller (3.2), is quadratically IO-FTS wrt $(\Omega, \mathcal{W}_2, Q(\cdot))$ with an $\mathcal{H}_\infty$ bound wrt $(\Gamma(\cdot), \Sigma(\cdot))$.

The resulting closed loop system has the following structure

$$\dot{x}(t) = (A_{cl}(t) + \Delta A_{cl}(t))\, x(t) + F(t)w(t) \tag{6.25a}$$

$$y(t) = C(t)x(t), \tag{6.25b}$$

where $A_{cl}(t) := A(t) + B(t)K(t)$, and

$$\Delta A_{cl}(t) = F_1(t)\Delta(t)(I - H(t)\Delta(t))^{-1}(E_1(t) + E_2(t)K(t)). \tag{$\diamond$}$$

The DLMIs/LMI feasibility problem in Theorem 6.1 can be numerically solved in an efficient way, as it has been shown in Section 6.2.2. However, when dealing with the controller design, in order to obtain a computationally tractable problem, it is necessary to pay the price of introducing some conservativeness. Indeed, when the closed loop system is considered, it is necessary to solve the DLMIs/LMI problem, stated in Theorem 6.1, by using a single matrix-valued optimization variable, i.e. by letting $P_1(\cdot) = P_2(\cdot)$. In order to do that, we first introduce the following result, which is a corollary of Theorem 6.1.

Corollary 6.1 Given the time interval Ω, and three continuous, positive definite matrix-valued functions $Q(\cdot)$, $\Gamma(\cdot)$, $\Sigma(\cdot)$, system (6.15) is quadratically IO finite-time stable wrt $(\Omega, \mathcal{W}_2, Q(\cdot))$ with an $\mathcal{H}_\infty$ bound wrt $(\Gamma(\cdot), \Sigma(\cdot))$, *if* there exists a piecewise continuously differentiable, positive definite matrix-valued function $P(\cdot)$ which satisfies, for $t \in \Omega$, the coupled DLMIs/LMI (the time argument is omitted for brevity)

$$\begin{pmatrix} \dot{P} + A^T P + PA & PF & PF_1 & E_1^T \\ F^T P & -R & 0 & 0 \\ F_1^T P & 0 & -I & H^T \\ E_1 & 0 & H & -I \end{pmatrix} < 0 \tag{6.26a}$$

$$\begin{pmatrix} \dot{P} + A^T P + PA + C^T \Sigma C & PF & PF_1 & E_1^T \\ F^T P & -\Gamma & 0 & 0 \\ F_1^T P & 0 & -I & H^T \\ E_1 & 0 & H & -I \end{pmatrix} < 0 \tag{6.26b}$$

$$P > C^T QC. \tag{6.26c}$$

▲

Proof: Let us assume that μ_1 and μ_2 are constant and equal to 1 in Theorem 6.1. Now, by letting

$$P_1(\cdot) = P_2(\cdot) =: P(\cdot),$$

inequality (6.17) reads

$$\begin{pmatrix} \dot{P} + A^T P + PA + L_1^T L_1 & P\Gamma & P\Gamma_1 + L_1^T \Pi \\ \Gamma^T P & -R & 0 \\ \Gamma_1^T P + H^T E_1 & 0 & -(I - H^T H) \end{pmatrix} < 0. \tag{6.27}$$

In order to show the equivalence between (6.27) and (6.26a) it is sufficient to apply Schur complements. Similar arguments can be exploited to show the equivalence between (6.21) and (6.26b), when a single optimization variable is considered. Finally, inequality (6.26c) is the same as (6.16c). ◇

When $R(\cdot) - \Gamma(\cdot)$ is nonnegative definite, the statement of Corollary 6.1 can be simplified, without introducing conservativeness. By following the same derivation which leads to (6.14), the following result readily follows.

Corollary 6.2 Assume that the matrix-function $R(\cdot) - \Gamma(\cdot)$ is nonnegative definite; then system (6.15) is quadratically IO finite-time stable wrt $(\Omega, \mathcal{W}_2, Q(\cdot))$ with an $\mathcal{H}_\infty$ bound wrt $(\Gamma(\cdot), \Sigma(\cdot))$, *if* there exists a piecewise continuously differentiable, positive definite matrix-valued function $P(\cdot)$ which satisfies, for all $t \in \Omega$, inequalities (6.26b) and (6.26c). ▲

We are now ready to introduce a pair of sufficient conditions to solve Problem 6.1.

Theorem 6.2 (Sufficient condition for Q-IO finite-time stabilization with an $\mathcal{H}_\infty$ bound [57]) Problem 6.1 admits a solution if there exist a piecewise continuously differentiable, positive definite matrix-valued function $\Pi(\cdot)$ and a piecewise continuous matrix-valued function $L(\cdot)$, such that the coupled DLMIs/LMI (the time argument is omitted for brevity)

$$\begin{pmatrix} -\dot{\Pi} + \Pi A^T + A\Pi + L^T B^T + BL & F & F_1 & \Pi E_1^T + L^T E_2^T \\ \cdots & -R & 0 & 0 \\ \cdots & \cdots & -I & H^T \\ \cdots & \cdots & \cdots & -I \end{pmatrix} < 0 \tag{6.28a}$$

$$\begin{pmatrix} -\dot{\Pi} + \Pi A^T + A\Pi + L^T B^T + BL & F & F_1 & \Pi E_1^T + L^T E_2^T & \Pi C^T \Sigma^{1/2} \\ \cdots & -\Gamma & 0 & 0 & 0 \\ \cdots & \cdots & -I & H^T & 0 \\ \cdots & \cdots & \cdots & -I & 0 \\ \cdots & \cdots & \cdots & \cdots & -I \end{pmatrix} < 0 \tag{6.28b}$$

$$\begin{pmatrix} \Pi & \Pi C^T \\ C\Pi & Q^{-1} \end{pmatrix} > 0. \tag{6.28c}$$

are satisfied[1] for all t in Ω. Moreover, a controller gain that solves Problem 6.1 is given by

$$K(t) = L(t)\Pi^{-1}(t).$$

▲

Proof: We apply the sufficient condition stated in Corollary 6.1 to the closed loop system given by the connection of system (6.2) and the controller (3.2). We obtain

$$\begin{pmatrix} \dot{P} + (A + BK)^T P + P(A + BK) & PF & PF_1 & E_1^T + K^T E_2^T \\ F^T P & -R & 0 & 0 \\ F_1^T P & 0 & -I & H^T \\ E_1 + E_2 K & 0 & H & -I \end{pmatrix} < 0 \tag{6.29}$$

Now, if we let $\Pi(\cdot) = P^{-1}(\cdot)$, and pre- and post-multiply inequality (6.29) by blockdiag $(\Pi(\cdot), I, I, I)$, condition (6.28a) is obtained by noticing that $\dot{\Pi}(t) = -\Pi(t)\dot{P}(t)\Pi(t)$, and letting $L(t) = K(t)\Pi(t)$.

By exploiting similar arguments it is possible to obtain condition (6.28b) from (6.26b); however, in this case, it is necessary to apply Schur complements to get rid of the non-linear term $\Pi(t)C^T(t)\Sigma(t)C(t)\Pi(t)$, that comes out when multiplying by $\Pi(\cdot)$ both sides of the term

$$\dot{P} + (A + BK)^T P + P(A + BK) + C^T \Sigma C.$$

Finally, by pre- and post-multiplying (6.26c) by $\Pi(\cdot)$, and by applying Schur complements, condition (6.28c) follows.

◇

In the same way the following simplified result can be proved, when $R(\cdot) - \Gamma(\cdot)$ is non-negative definite.

Theorem 6.3 Assume that $R(\cdot) - \Gamma(\cdot)$ is nonnegative definite; then Problem 6.1 admits a solution if there exist a piecewise continuously differentiable, positive definite matrix-valued function $\Pi(\cdot)$, and a piecewise continuous matrix-valued function $L(\cdot)$, such that (6.28b)-(6.28c) are satisfied. Moreover, a controller gain that solves Problem 6.1 is given by $K(\cdot) = L(\cdot)\Pi^{-1}(\cdot)$.

▲

Concluding, it is interesting to note that the issue considered in this chapter is related to another important problem, investigated in the control literature in recent years, namely the mixed $\mathcal{H}_\infty/\mathcal{L}_1$ problem.

To fix ideas and make a comparison between the two methodologies, let us assume that $R(\cdot) = \Gamma(\cdot)$; then, roughly speaking, given a signal in $\mathcal{L}_2$, the $\mathcal{H}_\infty$/FTS approach tries to minimize both the energy and the peak of the corresponding output; conversely the $\mathcal{H}_\infty/\mathcal{L}_1$ technique (see [118], and, more recently, [119]) for a given input in $\mathcal{L}_\infty$ ($\mathcal{L}_2$), tries to minimize the peak (energy) of the corresponding output. In other words, the mixed $\mathcal{H}_\infty/\mathcal{L}_1$ approach separately optimizes over two different classes of input signals, while the approach considered in this chapter performs a double optimization (peak and energy) over the same class of inputs.

---

1 For the sake of brevity, in (6.28a) and (6.28b) the lower triangular entries of the symmetric matrices on the left hand side have been replaced with dots.

Moreover, in [118, 119] the authors consider an infinite time-horizon control problem, which constraints the proposed approach to deal with time-invariant and asymptotically stable systems, while the $\mathcal{H}_\infty$/FTS methodology works over finite-time intervals, allows to treat linear time-varying systems and, in principle, does not require asymptotic stability of the system under consideration.

6.3.1 Numerical example: Cont'd

Let us consider again Problem 6.1 when the uncertain system introduced in Section 6.2.2 is considered. Let

$$B = \begin{pmatrix} 1 & 0 & 0 \end{pmatrix}^T$$
$$E_2 = 0.2.$$

In this case it is possible to design a state-feedback controller by solving the DLMIs/LMI optimization problem stated in Theorem 6.3 (since $R = \Gamma$), which admits a feasible solution. The structure for the optimization matrix-valued functions $\Pi(\cdot)$ and $L(\cdot)$ is, as usual, piecewise affine.

To check, for instance, the satisfaction of the robust IO-FTS requirement, Figure 6.3 reports the weighted output $|y(t)|_Q^2$ for 100 random realizations of the uncertain matrices, when the exogenous input $w(t) = 0.115$, with $t \in \Omega := [0, 15]$, is considered. As expected, $|y(t)|_Q^2$ is always less than 1. Furthermore, the $\mathcal{H}_\infty$ bound implies $\|y(\cdot)\|_{2,\Sigma} < 1$; in particular, for the considered 100 random simulations, it turns out that $\|y\|_{2,\Sigma} < 0.66$.

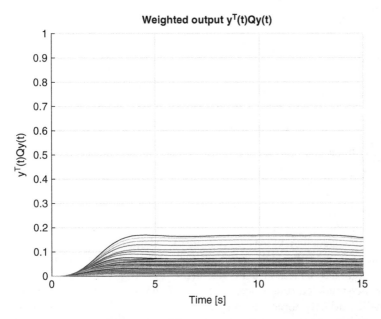

Weighted output $y^T(t)Qy(t)$

Figure 6.3 Weighted output $|y(t)|_Q^2$ for 100 random realizations of the closed loop uncertain system considered in Section 6.3.1.

6.4 Case study: Quadratic IO-FTS with an $\mathcal{H}_\infty$ Bound of the Inverted Pendulum

As a case study, in this section the inverted pendulum, shown in Figure E.4, is considered. Our goal is to design a state feedback controller that guarantees quadratic IO-FTS with an $\mathcal{H}_\infty$ bound in the time interval $\Omega = [0, 3]$, when the norm-bounded uncertainty modeled by

$$
\begin{aligned}
F_1 &= \begin{pmatrix} 0 & 0.9 & 0 & 0.9 \end{pmatrix}^T \\
E_1 &= \begin{pmatrix} 0 & 0.05 & 0.3 & 0 \end{pmatrix} \\
E_2 &= 0.012, \quad H = 0,
\end{aligned}
\tag{6.30}
$$

is considered.

Since we want to force disjoint constraints for the cart displacement and for the pendulum angle, we consider separately the two components of the output $y(\cdot)$, and we exploit the *structured* IO-FTS approach described in Chapter 5. To this aim, we consider the two outputs $y_1(t) = C_1 x(t) = s(t)$ and $y_2 = C_2 x(t) = \varphi(t)$, and we replace inequality (6.28c) with the following two inequalities

$$
\begin{pmatrix} \Pi(t) & \Pi(t)C_1^T(t) \\ C_1 \Pi(t) & q_1^{-1} \end{pmatrix} > 0, \quad \begin{pmatrix} \Pi(t) & \Pi(t)C_2^T \\ C_2 \Pi(t) & q_2^{-1} \end{pmatrix} > 0.
$$

Moreover, as we have already seen in Chapter 5, structured IO-FTS can be exploited also to limit the control input by adding to the feasibility problem in Theorem 6.2 the inequality

$$
\begin{pmatrix} \Pi(t) & L^T(t) \\ L(t) & v^{-1} \end{pmatrix} > 0,
$$

with $v > 0$.

By choosing

$$
\begin{aligned}
R &= 8, \quad q_1 = 1, \quad q_2 = 3.6 \\
v &= 10^{-4}, \quad \Gamma = 11, \quad \Sigma = \begin{pmatrix} 0.04 & 0 \\ 0 & 0.04 \end{pmatrix},
\end{aligned}
$$

the following bounds on s, φ and u are forced

$$
\begin{aligned}
\|s(\cdot)\|_\infty &< 1.05 \, m \\
\|\varphi(\cdot)\|_\infty &< 30° \\
\|u(\cdot)\|_\infty &< 100 \, N,
\end{aligned}
$$

when the exogenous input is such that $\|w\|_{2,R} \leq 1$. Note that a large bound on the control input is allowed, since the class of signals such that $\|w\|_{2,R} \leq 1$ includes also *impulsive* disturbances, that require a considerable control effort.

Figure 6.5 shows the simulated time traces of both $s(\cdot)$ and $\varphi(\cdot)$, and of the control input $u(\cdot)$, when the disturbance force depicted in Figure 6.4 is applied to the pendulum. The simulation has been carried out using the nonlinear model of the inverted pendulum.

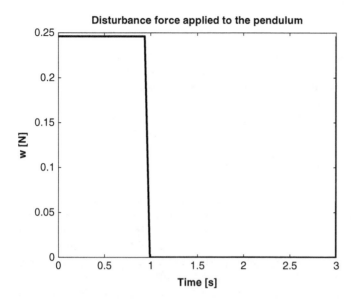

Figure 6.4 Disturbance force $w(\cdot)$ considered in the nonlinear simulation of the inverted pendulum.

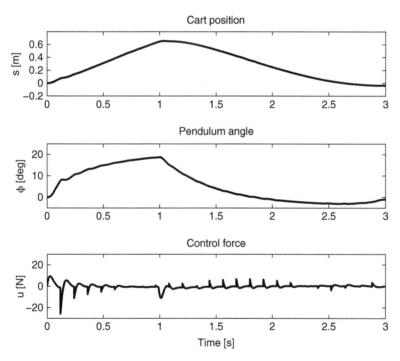

Figure 6.5 Time traces of the cart position s, of the pendulum angle φ, and of the control input u, when the disturbance shown in Figure 6.4 is applied to the inverted pendulum.

6.5 Summary

In this chapter the robustness problem in the IO-FTS setting has been considered. The concept of quadratic IO-FTS has been introduced, as the counterpart of the analogous property defined in the classical LS setting. Q-IO-FTS implies robust IO-FTS, namely IO-FTS for all admissible realizations of the uncertainty.

By mixing the properties of Q-IO-FTS with that one of $\mathcal{H}_\infty$ boundedness, we arrive to the definition of Q-IO-FTS with an $\mathcal{H}_\infty$ bound, that is useful when the goal is the minimization of the energy of the output signal on the interval Ω, in presence of all admissible exogenous inputs, and, at the same time, to constraint the *pointwise* value of the output itself.

A parallelism can be traced between the mixed $\mathcal{H}_\infty$/FTS control problem considered here, and the issue, investigated in the control literature in recent years, related to the mixed $\mathcal{H}_\infty/\mathcal{L}_1$ control problem [118, 119]. The main difference between the two approaches consists of the fact that the former performs a double optimization (peak and energy) over the same class of inputs, while the latter separately optimizes over two different classes of input signals ($\mathcal{L}_\infty$ and $\mathcal{L}_2$). Moreover, differently from the $\mathcal{H}_\infty/\mathcal{L}_1$ methodology, the $\mathcal{H}_\infty$/FTS control technique looks to finite-time intervals, and, in principle, does not require asymptotic stability of the system under consideration.

The first result of the chapter is a necessary and sufficient condition for Q-IO-FTS of a zero-input LTV system; the condition requires the solution of a feasibility problem constrained by a pair of DLMIs coupled with a time-varying LMI. Differently from the results derived in previous chapters, such condition requires the optimization over a pair of quadratic Lyapunov functions, namely $x^T(t)P_i(t)x(t)$, $i = 1, 2$; when, as usual, the analysis result is exploited to derive the design condition, in order to obtain computationally tractable conditions, we need to set $P_1(\cdot) = P_2(\cdot)$, so introducing a certain amount of conservativeness. Therefore the condition for the design of the state feedback controller turns out to be only sufficient.

Finally, the proposed technique has been illustrated by the design of a state feedback controller for the finite-time control of an inverted pendulum.

7

Impulsive Dynamical Linear Systems: IO-FTS Analysis

This chapter deals with the IO-FTS analysis of impulsive dynamical linear systems (IDLSs); IDLSs are useful to study the behavior of *discontinuous* dynamical systems, such as the rocking response of rigid blocks [120] and the automatic gear-box in cruise control (further examples can be found in [121]).

As for LTV systems, in this chapter we derive necessary and sufficient conditions for the IO-FTS of IDLSs when $\mathcal{W}_2$ exogenous inputs are considered, and sufficient conditions for IO-FTS when $\mathcal{W}_\infty$ signals are taken into account.

When $\mathcal{W}_2$ signals are dealt with, we shall use an approach based on linear operator theory and on the extension to impulsive systems of the concept of Reachability Gramian; the approach will lead to a condition based on a DLMI coupled to a difference LMI (D/DLMI); the DLMI is formally equivalent to the one obtained for LTV systems (see Theorem 2.3), with the difference that the initial value of the Lyapunov function, at each resetting time, is redefined by using the resetting condition given by the difference LMI. Finally, a third constraint is given by the usual time-varying LMI constraint, which accounts for the constraint on the output.

As a by-product of this result, we shall also prove an alternative necessary and sufficient condition for IO-FTS based on a coupled difference/differential Lyapunov equation (D/DLE). As in the case of LTV systems, the latter condition will be shown to be more efficient, from a computational point of view, with respect to the former D/DLMI/LMI condition, while the D/DLMI/LMI condition will be the starting point to derive computationally tractable conditions for the design of IO finite-time stabilizing controllers.

The latter result provided in the chapter involves a sufficient condition for IO-FTS in presence of $\mathcal{W}_\infty$ signals. Again, the statement of the theorem leads to a coupled D/DLMI, together with a time-varying LMI; a further LMI is needed to take into account the presence of the feedthrough term in the output equation.

Some numerical examples illustrate the application of the proposed methodology, together with the numerical issues regarding the conditions for IO-FTS in presence of $\mathcal{W}_2$ inputs.

The content of this and the next chapter is essentially based on the papers [122] and [123].

Finite-Time Stability: An Input-Output Approach, First Edition.
Francesco Amato, Gianmaria De Tommasi, and Alfredo Pironti.
© 2018 John Wiley & Sons Ltd. Published 2018 by John Wiley & Sons Ltd.

7.1 Background

We recall that the definition of IO-FTS for IDLSs is the same as the one for LTV systems, i.e., Definition 1.6. In the following, we shall present some preliminary results that will be exploited in Section 7.2 to prove the main result of this chapter.

7.1.1 Preliminary results for the W_2 case

When W_2 signals are considered, as usual we set $G(\cdot) = 0$ in (1.14), for well-posedness reasons.

The IDLS (1.14), with $G(\cdot) = 0$, can be regarded as a linear operator, say $\mathcal{ILS}$, that maps input signals into outputs; equipping the $\mathcal{L}_2(\Omega)$ and $\mathcal{L}_\infty(\Omega)$ spaces with the weighted norms $\| \cdot \|_{2,R}$ and $\| \cdot \|_{\infty,Q}$, respectively, the induced norm of the linear operator

$$\mathcal{ILS} : w(\cdot) \in \mathcal{L}_2(\Omega) \mapsto y(\cdot) \in \mathcal{L}_\infty(\Omega), \tag{7.1}$$

is given by

$$\|\mathcal{ILS}\|_{R,Q} = \sup_{\|w(\cdot)\|_{2,R}=1} \|y(\cdot)\|_{\infty,Q}, \tag{7.2}$$

that is the norm of the operator $\mathcal{ILS}$ is computed considering input signals that belong to W_2.

The next result readily follows from the definition of the operator norm given in (7.2).

Lemma 7.1 Given the time interval Ω, the class of input signals W_2, and the piecewise continuous, positive definite matrix-valued function $Q(\cdot)$, system (1.14), with $G(\cdot) = 0$, is IO-FTS wrt $(\Omega, W_2, Q(\cdot))$ *if and only if* $\|\mathcal{ILS}\|_{R,Q} < 1$. ▲

The following assumption is made on the resetting law of the IDLS (1.14).

Assumption 7.1 The matrix-valued function $J(\cdot)$ is such that $J(t_i)$ is non singular for each $t_i \in \mathcal{T}$. ◇

Under Assumption 7.1, it is possible to derive the following result, whose proof involves the concept of Reachability Gramian (see also Remark 7.2 below) and can be obtained by following the same guidelines of Theorem 2.2.

Lemma 7.2 Given the IDLS (1.14), the norm of the corresponding linear operator (7.1) is given by

$$\|\mathcal{ILS}\|_{R,Q} = \sup_{t \in \Omega} \lambda_{max}^{\frac{1}{2}} \left(Q^{\frac{1}{2}}(t)C(t)W(t)C^T(t)Q^{\frac{1}{2}}(t) \right), \tag{7.3}$$

for all $t \in \Omega$, where $\lambda_{max}(\cdot)$ denotes the maximum eigenvalue of the argument, and $W(\cdot)$ is the semidefinite positive matrix-valued function solution of the coupled D/DLE

$$\dot{W}(t) = A(t)W(t) + W(t)A^T(t) + F(t)R(t)^{-1}F^T(t), \quad t \notin \mathcal{T} \tag{7.4a}$$

$$W^+(t_i) = J(t_i)W(t_i)J^T(t_i), \quad t_i \in \mathcal{T} \tag{7.4b}$$

$$W(t_0) = 0. \tag{7.4c}$$

▲

The computational effort needed to check condition (7.3) is mainly the one related to the solution of the discontinuous differential equation (7.4), as discussed in the following remark. In the next sections it is shown that the condition in Lemma 7.2 can be used to check the IO-FTS of IDLSs in a computationally efficient way, especially when compared to the condition based on the D/DLMI feasibility problem (see Section 7.3 for more details).

Remark 7.1 Note that (7.4) actually defines $v + 1$ contiguous matrix DLEs; first, one solves (7.4a) in the interval $[t_0, t_1)$ starting from the initial condition provided by (7.4c). Then (7.4a) is solved in the interval $[t_1, t_2)$ starting from the initial condition returned by (7.4b), with $i = 1$, and so on for the following intervals. Therefore the solution $W(\cdot)$, defined over the whole interval Ω, coincides, in each interval $[t_i, t_{i+1})$, $i = 0, \dots, v$, with the (unique) solution of (7.4a) starting from $W^+(t_i)$ computed through (7.4c) and (7.4b). ◊

Remark 7.2 From the system-theoretic point of view, the matrix-valued function solution of (7.4) can be interpreted as the Reachability Gramian (see also [80] and [81]) of the IDLS obtained by (1.14), replacing $F(t)$ with $F(t)R(t)^{-1/2}$; in particular it is recursively defined as

$$W_r(t_0, t) := \int_{t_0}^{t} \Phi(t, \tau)F(\tau)R(\tau)^{-1}F^T(\tau)\Phi^T(t, \tau)d\tau, \quad t \in [t_0, t_1),$$

$$W_r^+(t_0, t_1) := J(t_1)W_r(t_0, t_1)J^T(t_1),$$

while, for $t \in [t_j, t_{j+1})$, $j = 1, \dots, v - 1$,

$$W_r(t_0, t) := \Phi(t, t_j)W_r^+(t_0, t_j)\Phi^T(t, t_j)$$
$$+ \int_{t_j}^{t} \Phi(t, \tau)F(\tau)R(\tau)^{-1}F^T(\tau)\Phi^T(t, \tau)d\tau,$$

$$W_r^+(t_0, t_{j+1}) := J(t_{j+1})W_r(t_0, t_{j+1})J^T(t_{j+1}),$$

and for $t \in [t_v, t_0 + T]$

$$W_r(t_0, t) := \Phi(t, t_v)W_r^+(t_0, t_v)\Phi^T(t, t_v)$$
$$+ \int_{t_v}^{t} \Phi(t, \tau)F(\tau)R(\tau)^{-1}F^T(\tau)\Phi^T(t, \tau)d\tau.$$

Once we introduce the Reachability Gramian, it is easy to verify that Assumption 7.1 is needed to guarantee the invertibility of the state- transition matrix, which implies the existence of $\Phi(\cdot, \cdot)$. ◊

7.2 Main Results: Necessary and Sufficient Conditions for IO-FTS in Presence of $\mathcal{W}_2$ Signals

In this section we provide a pair of necessary and sufficient conditions for the IO-FTS of the IDLS (1.14), with $G(\cdot) = 0$, in presence of $\mathcal{W}_2$ exogenous inputs.

We first introduce the following technical lemma, which is needed to prove the next theorem.

Lemma 7.3 Given $\epsilon > 0$, the solution of the coupled D/DLE

$$\dot{W}_\epsilon(t) = A(t)W_\epsilon(t) + W_\epsilon(t)A^T(t) + F(t)R(t)^{-1}F^T(t) + \epsilon I, \quad t \notin \mathcal{T} \tag{7.5a}$$
$$W_\epsilon^+(t_i) = J(t_i)W_\epsilon(t_i)J^T(t_i), \quad t_i \in \mathcal{T} \tag{7.5b}$$
$$W_\epsilon(t_0) = \epsilon I \tag{7.5c}$$

is the positive definite matrix-valued function

$$W_\epsilon(t) = W(t) + \epsilon\Phi(t, t_0)\Phi^T(t, t_0) + \epsilon \int_{t_0}^t \Phi(t, \tau)\Phi^T(t, \tau)d\tau, \tag{7.6}$$

where $W(\cdot)$ is the solution of the D/DLE (7.4). ▲

Proof: Let $t \in (t_i, t_{i+1}), i \in \{0, \dots, \nu\}$; taking the derivative with respect to t of both sides in (7.6), we have

$$\dot{W}_\epsilon(t) = \dot{W}(t) + \epsilon A(t)\Phi(t, t_0)\Phi^T(t, t_0) + \epsilon\Phi(t, t_0)\Phi^T(t, t_0)A^T(t)$$
$$+ \epsilon\left[A(t)\int_{t_0}^t \Phi(t, \tau)\Phi^T(t, \tau)d\tau + \int_{t_0}^t \Phi(t, \tau)\Phi^T(t, \tau)d\tau A^T(t) + \Phi(t, t)\Phi^T(t, t)\right]$$
$$= A(t)W(t) + W(t)A^T(t) + F(t)R(t)^{-1}F^T(t) + \epsilon I$$
$$+ A(t)\left[\epsilon\Phi(t, t_0)\Phi^T(t, t_0) + \epsilon \int_{t_0}^t \Phi(t, \tau)\Phi^T(t, \tau)d\tau\right]$$
$$+ \left[\epsilon\Phi(t, t_0)\Phi^T(t, t_0) + \epsilon \int_{t_0}^t \Phi(t, \tau)\Phi^T(t, \tau)d\tau\right]A^T(t). \tag{7.7}$$

From the last equality we see that $W_\epsilon(\cdot)$ satisfies (7.5a).
Now, evaluating in (7.6) the right limit in t_i of both sides, we have

$$W_\epsilon^+(t_i) = W^+(t_i) + \epsilon\Phi^+(t_i, t_0)\Phi^{+T}(t_i, t_0) + \epsilon \int_{t_0}^t \Phi^+(t_i, \tau)\Phi^{+T}(t_i, \tau)d\tau$$
$$= J(t_i)W(t_i)J^T(t_i) + \epsilon J(t_i)\Phi(t_i, t_0)\Phi^T(t_i, t_0)J^T(t_i)$$
$$+ \epsilon J(t_i)\int_{t_0}^{t_i} \Phi(t_i, \tau)\Phi^T(t_i, \tau)d\tau J^T(t_i). \tag{7.8}$$

Therefore $W_\epsilon(\cdot)$ satisfies (7.5b). Finally, from the definition of $W_\epsilon(\cdot)$ it readily follows that the initial condition (7.5c) is satisfied.

We are now ready to prove the following result.

Theorem 7.1 (Necessary and sufficient conditions for IO-FTS of IDLSs, $\mathcal{W}_2$ case [123]) Given the time interval Ω, the class of inputs $\mathcal{W}_2$, and the piecewise continuous, positive definite matrix-valued function $Q(\cdot)$, the following statements are equivalent:

i) The IDLS (1.14) is IO finite-time stable wrt $(\Omega, \mathcal{W}_2, Q(\cdot))$.

ii) The inequality

$$\sup_{t \in \Omega} \lambda_{\max} \left(Q^{\frac{1}{2}}(t) C(t) W(t) C^T(t) Q^{\frac{1}{2}}(t) \right) < 1 \tag{7.9}$$

holds, where $W(\cdot)$ is the positive semidefinite matrix function solution of the D/DLE (7.4).

iii) The coupled D/DLMI/LMI

$$\begin{pmatrix} \dot{P}(t) + A^T(t)P(t) + P(t)A(t) & P(t)F(t) \\ F^T(t)P(t) & -R(t) \end{pmatrix} < 0, \quad t \in \Omega, \quad t \notin \mathcal{T}, \tag{7.10a}$$

$$J^T(t_i)P^+(t_i)J(t_i) - P(t_i) < 0, \quad t_i \in \mathcal{T}, \tag{7.10b}$$

$$P(t) > C^T(t)Q(t)C(t), \quad t \in \Omega \tag{7.10c}$$

admit a positive definite solution $P(\cdot)$, which is piecewise continuously differentiable in each interval $[t_k, t_{k+1}), k = 0, \dots, \nu, t_{\nu+1} = t_0 + T$.

▲

Proof: We will prove the equivalence of the three statements by showing that **i)** ⇔ **ii)**, **ii)** ⇒ **iii)**, and **iii)** ⇒ **i)**.

[**i)** ⇔ **ii)**].

The equivalence between the two statements **i)** and **ii)** readily follows from Lemmas 7.1 and 7.2.

[**ii)** ⇒ **iii)**].

First, note that we have already proven that condition ii) implies the IO-FTS of the IDLS (1.14) wrt $(\Omega, \mathcal{W}_2, Q(\cdot))$; then, by continuity arguments, there exists a scalar $\eta > 0$ such that the IDLS given by (1.14a), (1.14c), and by the resetting law

$$x^+(t_i) = J(t_i)(1 + \eta) x(t_i), \quad t \in \mathcal{T}, \tag{7.11}$$

is IO finite-time stable wrt $(\Omega, \mathcal{W}_2, Q(\cdot))$. This circumstance, thanks again to the fact that i) ⇒ ii), implies that there exists a positive semidefinite matrix-valued solution $W_\eta(\cdot)$ to the coupled D/DLE

$$\dot{W}_\eta(t) = A(t)W_\eta(t) + W_\eta(t)A^T(t) + F(t)R(t)^{-1}F^T(t), \quad t \notin \mathcal{T} \tag{7.12a}$$

$$W_\eta^+(t_i) = (1 + \eta)^2 J(t_i)W_\eta(t_i)J^T(t_i), \quad t_i \in \mathcal{T} \tag{7.12b}$$

$$W_\eta(t_0) = 0, \tag{7.12c}$$

such that

$$\sup_{t \in \Omega} \lambda_{\max} \left(Q^{\frac{1}{2}}(t) C(t) W_\eta(t) C^T(t) Q^{\frac{1}{2}}(t) \right) < 1. \tag{7.13}$$

Now, given $\epsilon > 0$, let us apply Lemma 7.3 to the modified system given by (1.14a), (1.14c), and by the resetting law (7.11); let us denote by $W_{\eta,\epsilon}(\cdot)$ the positive definite solution of the D/DLE

$$\dot{W}_{\eta,\epsilon}(t) = A(t)W_{\eta,\epsilon}(t) + W_{\eta,\epsilon}(t)A^T(t) + F(t)R(t)^{-1}F^T(t) + \epsilon I, \quad t \notin \mathcal{T} \tag{7.14a}$$

$$W_{\eta,\epsilon}^+(t_i) = (1 + \eta)^2 J(t_i)W_{\eta,\epsilon}(t_i)J^T(t_i), \quad t_i \in \mathcal{T} \tag{7.14b}$$

$$W_{\eta,\epsilon}(t_0) = \epsilon I. \tag{7.14c}$$

From (7.14a) it follows that, for all $t \notin \mathcal{T}$, $W_{\eta,\epsilon}(\cdot)$ satisfies the DLMI

$$-\dot{W}_{\eta,\epsilon}(t) + A(t)W_{\eta,\epsilon}(t) + W_{\eta,\epsilon}(t)A^T(t) + F(t)R(t)^{-1}F^T(t) < 0. \tag{7.15}$$

Now letting

$$W_{\eta,\epsilon}(t) = P^{-1}(t), \tag{7.16}$$

it follows that $\dot{W}_{\eta,\epsilon}(t) = -P^{-1}(t)\dot{P}(t)P^{-1}(t)$, and inequality (7.15) reads

$$P^{-1}(t)\dot{P}(t)P^{-1}(t) + A(t)P^{-1}(t) + P^{-1}(t)A^T(t) + F(t)R^{-1}(t)F^T(t) < 0, \tag{7.17}$$

for all $t \in \Omega$ such that $t \notin \mathcal{T}$. By pre- and post-multiplying (7.17) by $P(t)$ we obtain

$$\dot{P}(t) + P(t)A(t) + A^T(t)P(t) + P(t)F(t)R^{-1}(t)F^T(t)P(t) < 0, \tag{7.18}$$

and (7.10a) readily follows by applying Schur complements.

Furthermore, from (7.14b) the following difference LMI holds for all t_i in $\mathcal{T}$

$$W_{\eta,\epsilon}^+(t_i) > J(t_i)W_{\eta,\epsilon}(t_i)J^T(t_i).$$

Taking into account (7.16), and by applying Schur complements, the previous inequality is equivalent to (7.10b).

Finally, to prove (7.10c), first note that $W_{\eta,\epsilon}(\cdot) \xrightarrow{\epsilon \to 0} W_\eta(\cdot)$; hence, from (7.13) and using continuity arguments, for the given η, there exists a sufficiently small ϵ such that

$$\sup_{t \in \Omega} \lambda_{\max}\left(Q^{\frac{1}{2}}(t)C(t)W_{\eta,\epsilon}(t)C^T(t)Q^{\frac{1}{2}}(t)\right) < 1. \tag{7.19}$$

Furthermore, from (7.16), condition (7.19) is equivalent to

$$I - Q^{\frac{1}{2}}(t)C(t)P^{-1}(t)C^T(t)Q^{\frac{1}{2}}(t) > 0, \tag{7.20}$$

for all t in Ω. By applying Schur complements, the last inequality reads

$$\begin{pmatrix} I & Q^{\frac{1}{2}}(t)C(t) \\ C^T(t)Q^{\frac{1}{2}}(t) & P(t) \end{pmatrix} > 0, \quad t \in \Omega. \tag{7.21}$$

From [96, Lemma 5.3] inequality (7.21) is equivalent to

$$\begin{pmatrix} P(t) & C^T(t)Q^{\frac{1}{2}}(t) \\ Q^{\frac{1}{2}}(t)C(t) & I \end{pmatrix} > 0, \quad t \in \Omega,$$

which yields (7.10c) by applying again Schur complements.

[iii) $\Rightarrow$ i)]. Let us consider the quadratic function $V(t,x) = x^T(t)P(t)x(t)$. Assuming that $t \notin \mathcal{T}$, the derivative with respect to time reads

$$\frac{d}{dt}(x^T(t)P(t)x(t)) = x^T(t)\dot{P}(t)x(t) + \dot{x}^T(t)P(t)x(t) + x^T(t)P(t)\dot{x}(t)$$
$$= x^T(t)(\dot{P}(t) + A^T(t)P(t) + P(t)A(t))x(t)$$
$$+ w^T(t)F^T(t)P(t)x(t) + x^T(t)P(t)F(t)w.$$

Condition (7.10a) implies that (the time argument is omitted for brevity)

$$\frac{d}{dt}(x^T P x) < w^T F^T P x + x^T P F w - x^T P F R^{-1} F^T P x.$$

Let $v = (R^{1/2}w - R^{-1/2}F^T Px)$, then

$$v^T v = w^T Rw + x^T PFR^{-1}F^T Px - w^T F^T Px - x^T PFw.$$

It follows that, for all $t \notin \mathcal{T}$,

$$\frac{d}{dt}(x^T(t)P(t)x(t)) < w^T(t)R(t)w(t) - v^T(t)v(t) < w^T(t)R(t)w(t). \tag{7.22}$$

First remember that $t_0 \notin \mathcal{T}$, then assume that in the time interval $]t_0, t]$ the state jumps h times, i.e.,

$$]t_0, t] \cap \mathcal{T} = \{t_1, t_2, \ldots, t_h\}.$$

Integrating (7.22) between t_0 and t_1, taking into account that $x(t_0) = 0$ we obtain

$$x(t_1)^T P(t_1)x(t_1) < \int_{t_0}^{t_1} w^T(\sigma)Rw(\sigma)d\sigma. \tag{7.23a}$$

Similarly we obtain

$$x(t_2)^T P(t_2)x(t_2) - x^+(t_1)^T P^+(t_1)x^+(t_1) < \int_{t_1}^{t_2} w^T(\sigma)Rw(\sigma)d\sigma \tag{7.23b}$$

$$\ldots$$

$$x(t_h)^T P(t_h)x(t_h) - x^+(t_{h-1})^T P^+(t_{h-1})x^+(t_{h-1}) < \int_{t_{h-1}}^{t_h} w^T(\sigma)Rw(\sigma)d\sigma \tag{7.23c}$$

$$x(t)^T P(t)x(t) - x^+(t_h)^T P^+(t_h)x^+(t_h) < \int_{t_h}^{t} w^T(\sigma)Rw(\sigma)d\sigma, \tag{7.23d}$$

with $t \leq t_0 + T$. From (7.23) it readily follows that

$$x^T(t)P(t)x(t)$$

$$+ \sum_{i=1}^{h}(x^T(t_i)P(t_i)x(t_i) - x^+(t_i)^T P^+(t_i)x^+(t_i)) < \int_{t_0}^{t} w^T(\sigma)Rw(\sigma)d\sigma. \tag{7.24}$$

Since

$$x^T(t_k)P(t_k)x(t_k) - x^+(t_k)^T P^+(t_k)x^+(t_k) = x^T(t_k)P(t_k)x(t_k)$$
$$- x^T(t_k)J^T(t_k)P^+(t_k)J(t_k)x(t_k), \tag{7.25}$$

for all $t_k \in \mathcal{T}$, condition (7.10b) implies that

$$\sum_{i=1}^{h}(x^T(t_i)P(t_i)x(t_i) - x^+(t_i)^T P^+(t_i)x^+(t_i)) > 0, \tag{7.26}$$

hence, taking into account that $w(\cdot)$ belongs to $\mathcal{W}_2$, it follows

$$x^T(t)P(t)x(t) < \int_{t_0}^{t} w^T(\sigma)Rw(\sigma) \leq \|w\|_{2,R}^2 \leq 1.$$

Exploiting the terminal condition (7.10c), we obtain

$$y(t)^T Q(t)y(t) = x^T(t)C^T(t)Q(t)C(t)x(t) \leq x^T(t)P(t)x(t) < 1,$$

for all $t \in \Omega$; from this the proof follows. $\qquad \diamond$

Remark 7.3   Note that the D/DLE and D/DLMI conditions given in Theorem 7.1 hold also for the SLSs introduced at the end of Section 1.4.2.

Furthermore, Theorem 7.1 holds also when the family of the SLS is composed of dynamics of different order. In this case the D/DLE (D/DLMI) can be written by choosing $v + 1$ matrix-valued functions $W_i(\cdot), (P_i(\cdot))$, with the same dimension of $A_{\sigma(t)}(\cdot)$ with $t \in [t_{i-1}, t_i), t_i \in \mathcal{T}$.   ◇

7.3   Example and Computational Issues

In this section we discuss some computational issues related to the solution of the D/DLE (7.4) and the D/DLMI/LMI (7.10), by means of a numerical example.

As expected, the example shows that condition ii) in Theorem 7.1, which is based on the solution of the D/DLE (7.4), is much more efficient, from the computational point of view, than the D/DLMI/LMI (7.10).

Let us consider the time-varying IDLS

$$A = \begin{pmatrix} -2.5 + 0.2 \cdot t & -6.3 \\ 4 & 0.2 \cdot t \end{pmatrix}, \quad F = \begin{pmatrix} 2 \\ 0 \end{pmatrix} \tag{7.27a}$$

$$C = \begin{pmatrix} 1.2 & 3.2 \end{pmatrix}, \quad J = \begin{pmatrix} 1.1 & 0 \\ 0 & -0.8 \end{pmatrix}. \tag{7.27b}$$

The time interval we consider in this example is $\Omega = [0, 2]$, while the resetting times are

$$\mathcal{T} = \{0.25, 0.5, 0.75, 1, 1.5, 1.8\}.$$

The input weighting matrix is taken constant and equal to

$$R = 0.7.$$

Figure 7.1 shows an example of time response of the IDLS (7.27) in the interval Ω when an input in $\mathcal{W}_2$ is considered. For this time response, Figure 7.2 depicts the time behavior of the weighted output, when Q is taken equal to 2. It clearly shows that, making this choice for Q, the system is not IO-FTS wrt $(\Omega, \mathcal{W}_2, Q)$.

Now we want to exploit the results of Theorem 7.1 to estimate, by means of a linear search, the maximum value of Q, say Q_{max}, such that system (7.27) is IO finite-time stable wrt $(\Omega, \mathcal{W}_2, Q_{max})$.

Taking Q equal to Q_{max}, for a fixed R, corresponds to the *minimization* of the maximum output in the time interval Ω, when the class of inputs $\mathcal{W}_2$ is applied to the IDLS (7.27).

Given a matrix-valued function $P(\cdot)$, with the usual piecewise affine structure, according to Appendix C.2, it is then possible to exploit standard optimization tools, such as SeDuMi [91] or the MATLAB LMI Toolbox® [90], to find a solution that verifies condition iii) in Theorem 7.1.

Tables 7.1 and 7.2 present the estimates of Q_{max}, the corresponding values of the *sampling time* T_s, and the computation time for a single iteration of the linear search, obtained using the Matlab LMI toolbox on a PC equipped with a 3.4 GHz Intel Core i7 processor with 8 GB of RAM. Such results clearly show that condition ii), which is based on the solution of the D/DLE, is computationally more efficient, since it permits to achieve a better estimation for Q_{max}, without requiring a big computational effort.

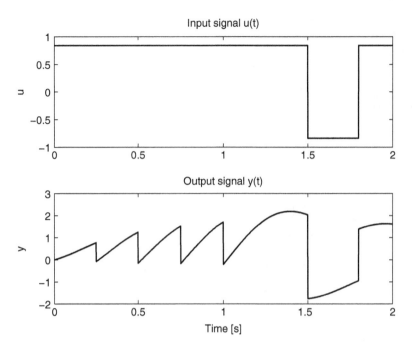

Figure 7.1 Example of the time behavior of the impulsive system (7.27), when an input in $\mathcal{W}_2$ is considered.

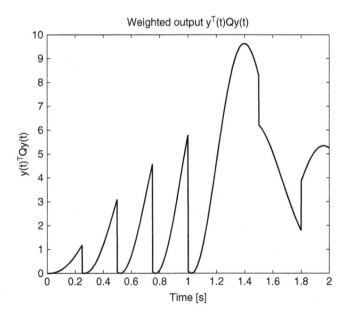

Figure 7.2 Weighted output $|y(t)|_Q^2$ of system (7.27) when the input shown in Figure 7.1 is applied to the impulsive system (7.27), and Q is taken equal to 2.

Table 7.1 Values of Q_{max} obtained exploiting condition ii) in Theorem 7.1 for the IDLS system (7.27).

Sample Time (T_s) [ms]	Q_{max}	Computation time for the solution of the D/DLE (7.4) [s]
10	0.0900	0.19
1	0.0910	0.22
0.1	0.0918	0.7

Table 7.2 Values of Q_{max} obtained exploiting condition iii) in Theorem 7.1 for the IDLS system (7.27).

Sample Time (T_s) [ms]	Q_{max}	Average computation time for a single iteration [s]
50	0.0740	2.6
25	0.0796	14.2
10	0.0837	298.4

On the other hand, as said before, condition iii) will be the starting point to derive design conditions for IO finite-time stabilization.

7.4 Main Result: A Sufficient Condition for IO-FTS in Presence of $\mathcal{W}_\infty$ Signals

In this section we assume that $G(\cdot) \neq 0$, and consider the IO-FTS problem for the IDLS (1.14) in presence of $\mathcal{W}_\infty$ signals. The following sufficient condition for IO-FTS is stated here for the first time.

Theorem 7.2 (Sufficient condition for IO-FTS of IDLSs, $\mathcal{W}_\infty$ case) Given the time interval Ω, the class of inputs $\mathcal{W}_\infty$, and the piecewise continuous, positive definite matrix-valued function $Q(\cdot)$, the IDLS (1.14) is IO finite-time stable wrt $(\Omega, \mathcal{W}_\infty, Q(\cdot))$, if there exists a scalar function $\theta(\cdot)$, $\theta(t) > 1$, $t \in \Omega$, such that the coupled D/DLMI/LMIs

$$\begin{pmatrix} \dot{P}(t) + A^T(t)P(t) + P(t)A(t) & P(t)F(t) \\ F^T(t)P(t) & -R(t) \end{pmatrix} < 0, \quad t \notin \mathcal{T} \tag{7.28a}$$

$$J^T(t_i)P^+(t_i)J(t_i) - P(t_i) < 0, \quad t_i \in \mathcal{T} \tag{7.28b}$$

$$\theta(t)R(t) - R(t) > 2\theta(t)G^T(t)Q(t)G(t), \quad t \in \Omega \tag{7.28c}$$

$$P(t) > 2\theta(t)C^T(t)\tilde{Q}(t)C(t), \quad t \in \Omega \tag{7.28d}$$

where $\tilde{Q}(t) = (t - t_0)Q(t)$, admit a positive definite solution $P(\cdot)$, which is piecewise continuously differentiable in each interval $[t_k, t_{k+1})$, $k = 0, \ldots, v$, $t_{v+1} = t_0 + T$. ▲

Proof: Let us consider the quadratic function $V(t,x) = x^T(t)P(t)x(t)$; by following the same guidelines of the proof that condition iii) implies condition i) in Theorem 7.1, we obtain that, for all $t \in \Omega/\mathcal{T}$,

$$\frac{d}{dt}(x^T(t)P(t)x(t)) < w^T(t)R(t)w(t) \leq 1, \tag{7.29}$$

where we have used the fact that $\|w\|_{\infty,R} \leq 1$.

Following the same steps as in (7.23)–(7.25), and using (7.26), it follows that, for all $t \in \Omega$,

$$x^T(t)P(t)x(t) < t - t_0. \tag{7.30}$$

Now, as in the proof of Theorem 2.4, we obtain that

$$y^T(t)Q(t)y(t) \leq 2(x^T(t)C^T(t)Q(t)C(t)x(t) + w^T(t)G^T(t)Q(t)G(t)w(t)). \tag{7.31}$$

Exploiting the fact that $\|w\|_{\infty,R} \leq 1$, condition (7.31), together with conditions (7.28c), (7.28d), and (7.30), imply that

$$y^T(t)Q(t)y(t) < \frac{1}{\theta(t)}\frac{x^T(t)P(t)x(t)}{t - t_0} + \frac{\theta(t) - 1}{\theta(t)}$$

$$< \frac{1}{\theta(t)} + \frac{\theta(t) - 1}{\theta(t)} = 1; \tag{7.32}$$

from the last inequality the proof follows. ◇

7.4.1 An illustrative example

Let us consider the second-order IDLS with the continuous-time dynamic defined by

$$A = \begin{pmatrix} -2.5 + 0.5t & -6.25 \\ 4 & 0.5t \end{pmatrix}, \qquad F = \begin{pmatrix} 2 \\ 0 \end{pmatrix} \tag{7.33a}$$

$$C = \begin{pmatrix} 0 & 3.125 \end{pmatrix}, \qquad G = 0, \tag{7.33b}$$

with the resetting law defined by

$$J = \begin{pmatrix} -0.8 & 0 \\ 0 & -0.8 \end{pmatrix}. \tag{7.34}$$

Given the resetting times set

$$\mathcal{T} = \{0.25, 0.5, 0.75, 1, 1.25, 1.5, 1.75\}, \tag{7.35}$$

and letting

$$R = 1, \qquad Q = 0.1, \qquad \Omega := [0, 2],$$

we exploit the results given in Theorem 7.2 to check IO-FTS of the given IDLS wrt $(\Omega, \mathcal{W}_\infty, Q)$.

In order to recast the coupled D/DLMI/LMIs condition in Theorem 7.2 in terms of LMIs, the matrix-valued function $P(\cdot)$ has been assumed piecewise affine with jumps in

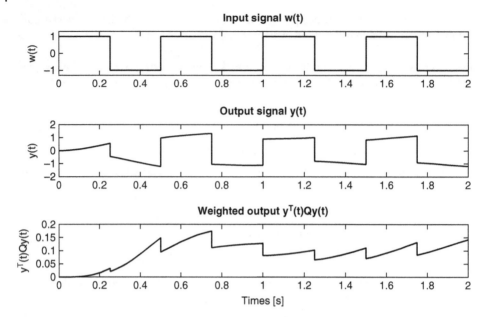

Figure 7.3 Time evolution of the exogenous input $w(\cdot)$, of the output $y(\cdot)$, and of the weighted output $|y(t)|_Q^2$ for the IDLS considered in Section 7.4.1.

correspondence of the resetting times. Then, the procedure described in Appendix C.2 has been used.

Exploiting standard optimization tools such as the MATLAB LMI Toolbox® ([90]), it is possible to find a matrix function $P(\cdot)$ that verifies the conditions provided in Theorem 7.2 and has the structure described above. Hence the considered IDLS is IO finite-time stable wrt $(\Omega, \mathcal{W}_\infty, Q)$.

Figure 7.3 shows the time evolution of the exogenous input $w(t)$, of the output signal $y(t)$, and of the weighted output $|y(t)|_Q^2$. It is important to note that the considered input represents the worst input in the class $\mathcal{W}_\infty$, given the resetting law (7.34) and the resetting times set (7.35).

7.5 Summary

In this chapter we have investigated the IO-FTS problem for the class of IDLSs, in presence of both $\mathcal{W}_2$ and $\mathcal{W}_\infty$ exogenous inputs.

When $\mathcal{W}_2$ signals are considered, a result formally similar to that one found for LTV systems has been proven, thanks to the extension of the Reachability Gramian theory to the class of IDLSs. Indeed, the main result of the chapter, namely Theorem 7.1, is a necessary and sufficient condition for the IO-FTS of system (1.14), in terms of both a DLE based condition, and a DLMI based condition respectively.

As it was already shown in Chapter 2, the DLE-based condition is much more efficient from the computational point of view; however, the DLMI condition will represent the starting point for the solution of the design problem, tackled in the next chapter.

As for the IO-FTS in presence of $\mathcal{W}_\infty$ signals, a sufficient condition has been stated, which is again formally equivalent to that one of the LTV system case, with a further set of LMIs to take into account the switching dynamics. Again, a numerical example has illustrated the application of this result.

8

Impulsive Dynamical Linear Systems: IO Finite-Time Stabilization via Dynamical Controllers

In this chapter, we face the design problem for the IDLSs investigated in Chapter 7.

By using the analysis conditions stated in Theorems 7.1 and 7.2, we shall provide a necessary and sufficient condition, as well as a sufficient condition for the existence of a dynamical output feedback controller that IO finite-time stabilizes the closed loop system with respect to $\mathcal{W}_2$ and $\mathcal{W}_\infty$ signals.

Then, as a particular case of the output feedback design, a pair of conditions for the existence of finite-time stabilizing state feedback controllers are derived.

As usual, all the involved conditions will require the solution of a feasibility problem constrained by coupled D/DLMI/LMI conditions.

8.1 Problem Statement

The synthesis problem we deal with in this chapter is summarized as follows. For the sake of simplicity, we assume that $G(\cdot) = 0$ both in the $\mathcal{W}_2$ and $\mathcal{W}_\infty$ cases.

Problem 8.1 **(IO finite-time stabilization of IDLSs)** Consider the IDLS

$$\dot{x}(t) = A(t)x(t) + B(t)u(t) + F(t)w(t), \quad x(t_0) = 0, \quad t \in \Omega, \quad t \notin \mathcal{T} \tag{8.1a}$$

$$x^+(t_i) = J(t_i)x(t_i), \quad t_i \in \mathcal{T} \tag{8.1b}$$

$$y(t) = C(t)x(t), \quad t \in \Omega, \tag{8.1c}$$

where $u(\cdot)$ is the control input, $w(\cdot)$ is the exogenous input belonging to a given class of signals $\mathcal{W}$, and $\mathcal{T} = \{t_1, \dots, t_\nu\}$ is the resetting times set, with $t_1 > t_0$, defined in Section 1.4.2. Given the piecewise continuous, positive definite matrix-valued function $Q(\cdot)$, defined in Ω, find a dynamic output feedback controller

$$\dot{x}_c(t) = A_K(t)x_c(t) + B_K y(t), \quad x_c(t_0) = 0, \quad t \in \Omega, \quad t \notin \mathcal{T} \tag{8.2a}$$

$$x_c^+(t) = J_K(t_i)x_c(t_i) + H_K(t_i)y(t_i), \quad t_i \in \mathcal{T} \tag{8.2b}$$

$$u(t) = C_K(t)x_c(t) + D_K(t)y(t), \quad t \in \Omega, \tag{8.2c}$$

Finite-Time Stability: An Input-Output Approach, First Edition.
Francesco Amato, Gianmaria De Tommasi, and Alfredo Pironti.
© 2018 John Wiley & Sons Ltd. Published 2018 by John Wiley & Sons Ltd.

where $x_K(\cdot)$ has the same dimension of $x(\cdot)$, and the controller matrices are piecewise continuous and have compatible dimensions, such that the closed-loop system

$$\begin{pmatrix} \dot{x}(t) \\ \dot{x}_c(t) \end{pmatrix} = \begin{pmatrix} A(t) + B(t)D_K(t)C(t) & B(t)C_K(t) \\ B_K(t)C(t) & A_K(t) \end{pmatrix} \begin{pmatrix} x(t) \\ x_c(t) \end{pmatrix} + \begin{pmatrix} F(t) \\ 0 \end{pmatrix} w(t)$$

$$=: A_{CL}(t)x_{CL}(t) + F_{CL}(t)w(t), \quad t \in \Omega, \quad t \notin \mathcal{T} \tag{8.3a}$$

$$\begin{pmatrix} x^+(t_i) \\ x_c^+(t_i) \end{pmatrix} = \begin{pmatrix} J(t_i) & 0 \\ H_K(t_i)C(t_i) & J_K(t_i) \end{pmatrix} \begin{pmatrix} x(t_i) \\ x_c(t_i) \end{pmatrix}$$

$$=: J_{CL}(t_i)x_{CL}(t_i), \quad t_i \in \mathcal{T} \tag{8.3b}$$

$$y(t) = \begin{pmatrix} C(t) & 0 \end{pmatrix} \begin{pmatrix} x(t) \\ x_c(t) \end{pmatrix}$$

$$=: C_{CL}(t)x_{CL}(t), \quad t \in \Omega, \tag{8.3c}$$

is IO-FTS wrt $(\mathcal{W}, Q(\cdot), \Omega)$. ◇

In the following section we shall provide a necessary and sufficient condition for the existence of a controller in the form (8.2), such that Problem 8.1 admits a feasible solution in presence of $\mathcal{W}_2$ signals.

Finally, in Section 8.3 we shall prove a sufficient condition for the existence of a controller in the form (8.2), such that Problem 8.1 admits a feasible solution in presence of $\mathcal{W}_\infty$ signals. As a particular case of the above results, conditions for the resolvability of the state feedback problem will be provided.

8.2 IO Finite-Time Stabilization of IDLSs: $\mathcal{W}_2$ Signals

In this section we exploit the D/DLMI feasibility problem introduced in condition **iii)** of Theorem 7.1, to derive a necessary and sufficient condition for IO finite-time stabilization of IDLS via output feedback, in presence of $\mathcal{W}_2$ exogenous inputs.

Theorem 8.1 (Necessary and sufficient condition for IO finite-time stabilization of IDLSs via output feedback; $\mathcal{W}_2$ case [123]) Problem 8.1 is solvable, in presence of $\mathcal{W}_2$ signals, *if and only if* there exist two symmetric matrix-valued functions $T(\cdot)$ and $S(\cdot)$ that are piecewise continuously differentiable in each interval $[t_k, t_{k+1})$, $k = 0, \ldots, v$, $t_{v+1} = t_0 + T$, four piecewise continuous matrix-valued functions $\hat{A}_K(\cdot)$, $\hat{B}_K(\cdot)$, $\hat{C}_K(\cdot)$, $D_K(\cdot)$, and $2v$ matrices $\hat{J}_{K_i}$ and $\hat{H}_{K_i}$, with $i = 1, \ldots, v$, such that the following coupled D/DLMI/LMI are satisfied

$$\begin{pmatrix} \Theta_{11}(t) & \Theta_{12}(t) & 0 \\ \Theta_{12}^T(t) & \Theta_{22}(t) & T(t)F(t) \\ 0 & F^T(t)T(t) & -R(t) \end{pmatrix} < 0, \quad t \in \Omega, \quad t \notin \mathcal{T} \tag{8.4a}$$

$$\begin{pmatrix} \Xi_{11}(t_i) & \Xi_{12}(t_i) \\ \Xi_{12}^T(t_i) & \Xi_{22}(t_i) \end{pmatrix} < 0, \quad t_i \in \mathcal{T} \tag{8.4b}$$

$$\begin{pmatrix} \Psi_{11}(t) & \Psi_{12}(t) & 0 \\ \Psi_{12}^T(t) & S(t) & S(t)C^T(t) \\ 0 & C(t)S(t) & Q^{-1}(t) \end{pmatrix} > 0, \quad t \in \Omega \tag{8.4c}$$

where

$$\begin{aligned} \Theta_{11}(t) &= -\dot{S}(t) + A(t)S(t) + S(t)A^T(t) + B(t)\hat{C}_K(t) \\ &\quad + \hat{C}_K^T(t)B^T(t) + F(t)R^{-1}(t)F^T(t), \\ \Theta_{12}(t) &= A(t) + \hat{A}_K^T(t) + B(t)D_K(t)C(t) + F(t)R^{-1}(t)F^T(t)T(t), \\ \Theta_{22}(t) &= \dot{T}(t) + T(t)A(t) + A^T(t)T(t) + \hat{B}_K(t)C(t) + C^T(t)\hat{B}_K^T(t), \\ \Xi_{11}(t_i) &= -\begin{pmatrix} S(t_i) & I \\ I & T(t_i) \end{pmatrix}, \\ \Xi_{12}(t_i) &= \begin{pmatrix} S(t_i)J^T(t_i) & \hat{J}_{K_i}^T \\ J^T(t_i) & J^T(t_i)T^+(t_i) + C^T(t_i)\hat{H}_{K_i}^T \end{pmatrix}, \\ \Xi_{22}(t_i) &= -\begin{pmatrix} S^+(t_i) & I \\ I & T^+(t_i) \end{pmatrix}, \\ \Psi_{11}(t) &= T(t) - C^T(t)Q(t)C(t), \\ \Psi_{12}(t) &= I - C^T(t)Q(t)C(t)S(t). \end{aligned}$$

▲

Proof: From Theorem 7.1 it follows that the IDLS (8.3) is IO-FTS wrt $(\mathcal{W}_2, Q(\cdot), \Omega)$ if and only if there exists a positive definite matrix-valued function $P(\cdot)$, of compatible dimensions, such that

$$\dot{P}(t) + A_{CL}^T(t)P(t) + P(t)A_{CL}(t) + P(t)F_{CL}(t)R(t)^{-1}F_{CL}^T(t)P(t) < 0, \quad t \notin \mathcal{T}, \tag{8.5a}$$

$$J_{CL}^T(t_i)P^+(t_i)J_{CL}(t_i) - P(t_i) < 0, \quad t_i \in \mathcal{T}, \tag{8.5b}$$

$$P(t) > C_{CL}^T(t)Q(t)C_{CL}(t), \quad t \in \Omega. \tag{8.5c}$$

By Defining $P(\cdot)$ as in (3.14), $\Pi_i(\cdot)$, $i = 1, 2$, as in (3.15), and following the same steps as in (3.16)–(3.17), the equivalence between (8.5a) and (8.4a), and between (8.5c) and (8.4c) follows.

In order to show the equivalence between (8.5b) and (8.4b), first note that, applying Schur complements arguments, (8.5b) can be rewritten as

$$\begin{pmatrix} -P(t_i) & J_{CL}^T(t_i)P^+(t_i) \\ P^+(t_i)J_{CL}(t_i) & -P^+(t_i) \end{pmatrix} < 0, \quad t_i \in \mathcal{T}. \tag{8.6}$$

By pre- and post-multiplying (8.6) by $diag(\Pi_1(t_i), \Pi_1^+(t_i))^T$ and $diag(\Pi_1(t_i), \Pi_1^+(t_i))$, respectively, we obtain[1]

---

[1] In the following inequalities the time dependence is discarded in order to avoid awkward notation; in particular, given a matrix function $H(\cdot)$, we let $H(t_i) = H$, and $H^+(t_i) = H^+$.

$$\begin{pmatrix} S & N & 0 & 0 \\ I & 0 & 0 & 0 \\ 0 & 0 & S^+ & N^+ \\ 0 & 0 & I & 0 \end{pmatrix} \begin{pmatrix} -T & -M & J^T T^+ + C^T H_K^T M^{+T} & J^T M^+ + C^T H_K^T U^+ \\ -M^T & -U & J_K^T M^{+T} & J_K^T U^+ \\ T^+ J + M^+ H_K C & M^+ J_K & -T^+ & -M^+ \\ M^{+T} J + U^+ H_K C & U^+ J_K & -M^{+T} & -U^+ \end{pmatrix}$$

$$\times \begin{pmatrix} S & I & 0 & 0 \\ N^T & 0 & 0 & 0 \\ 0 & 0 & S^+ & I \\ 0 & 0 & N^{+T} & 0 \end{pmatrix} < 0,$$

for all $t_i \in \mathcal{T}$, which, taking into account (3.16), turns to be equivalent to

$$\begin{pmatrix} -S & -I & SJ^T & SJ^T T^+ + SC^T H_K^T M^{+T} + NJ_K^T M^{+T} \\ -I & -T & J^T & J^T T^+ + C^T H_K^T M^{+T} \\ JS & J & -S^+ & -I \\ T^+ JS + M^+ H_K CS + M^+ J_K N^T & T^+ J + M^+ H_K C & -I & -T^+ \end{pmatrix} < 0,$$

(8.7)

for all $t_i \in \mathcal{T}$.

Eventually, condition (8.4b) follows from (8.7) once we let

$$\hat{J}_{K_i}(t_i) := M^+(t_i)J_K(t_i)N(t_i)^T + M^+(t_i)H_K(t_i)C(t_i)S(t_i) + T^+(t_i)J(t_i)S(t_i) \qquad (8.8a)$$

$$\hat{H}_{K_i}(t_i) := M^+(t_i)H_K(t_i), \qquad (8.8b)$$

for all $t_i \in \mathcal{T}$. ◇

Although Theorem 8.1 provides a necessary and sufficient condition to solve Problem 8.1, when $\mathcal{W}_2$ exogenous disturbances are considered, the solution of the D/DLMI feasibility problem does not directly allows to compute all the controller matrices. In order to obtain these matrices, starting from the solution of the D/DLMI problem, the following procedure can be applied.

Procedure 8.1 Assuming that the hypotheses of Theorem 8.1 are satisfied, in order to design the controller, the following steps have to be followed:

i) Find $T(\cdot)$, $S(\cdot)$, $\hat{A}_K(\cdot)$, $\hat{B}_K(\cdot)$, $\hat{C}_K(\cdot)$, $D_K(\cdot)$, $\hat{J}_{K_i}$, and $\hat{H}_{K_i}$, $i = 1, \ldots, v$, such that (8.4) are satisfied.

ii) Let $N(\cdot)$ be any nonsingular matrix-valued function (e.g., $N(t) = I$ for all $t \in \Omega$), and let

$$M(t) = [I - T(t)S(t)]N^{-T}(t).$$

iii) Obtain $A_K(\cdot)$, $B_K(\cdot)$ and $C_K(\cdot)$ by inverting (3.17), and $J_K(t_i)$ and $H_K(t_i)$ from (8.8).

◇

To conclude this section we now introduce the next result, which provides a necessary and sufficient condition for the IO finite-time stabilization of IDLS via state feedback, i.e., when $u(t) = K(t)x(t)$.

Theorem 8.2 (Necessary and sufficient condition for IO finite-time stabilization of IDLSs via state feedback; W_2 case [123]) Problem 8.1 , with $C(\cdot) = I$, in presence of W_2 signals, is solvable via state feedback control *if and only if* there exist a positive definite matrix-valued function $\Upsilon(\cdot)$, which is piecewise continuously differentiable in each interval $[t_k, t_{k+1})$, $k = 0, \dots, \nu$, $t_{\nu+1} = t_0 + T$, and a piecewise continuous matrix-valued function $L(\cdot)$, such that

$$\begin{pmatrix} -\dot{\Upsilon}(t) + A(t)\Upsilon(t) + \Upsilon(t)A(t)^T + L(t)^T B(t)^T + B(t)L(t) & F(t) \\ F^T(t) & -R(t) \end{pmatrix} < 0,$$
$$t \notin \mathcal{T}, \qquad (8.9a)$$

$$\begin{pmatrix} -\Upsilon^+(t_i) & J(t_i)\Upsilon(t_i) \\ \Upsilon(t_i)J(t_i) & -\Upsilon(t_i) \end{pmatrix} < 0, \quad t_i \in \mathcal{T} \qquad (8.9b)$$

$$\begin{pmatrix} \Upsilon(t) & \Upsilon(t)C^T(t) \\ C(t)\Upsilon(t) & Q^{-1}(t) \end{pmatrix} > 0, \quad \forall \, t \in \Omega. \qquad (8.9c)$$

In this case a controller gain that solves Problem 8.1 via state feedback is

$$K(t) = L(t)\Upsilon^{-1}(t), \quad t \in \Omega.$$

▲

Proof: The theorem can be easily proven by replacing $A(t)$ by $A(t) + B(t)K(t)$ in (7.10a), by pre- and post-multiplying condition (7.10a) by

$$\text{diag}\,(P^{-1}(t), P^{-1}(t)) =: \text{diag}\,(\Upsilon(t), \Upsilon(t)),$$

and by pre- and post-multiplying conditions (7.10b)-(7.10c) by $\Upsilon(t)$, letting $L(t) = K(t)\Upsilon(t)$, and finally using Schur complements. ◇

8.2.1 A numerical example

In order to show the effectiveness of the proposed approach, we exploit the coupled D/DLMI/LMI-based feasibility problem from Theorem 8.1 to design an output feedback finite-time stabilizing controller for the third-order IDLS

$$A = \begin{pmatrix} 0 & 1 & 0 \\ 0 & 0 & 1 \\ -0.02 & -0.08 & 1.1 \end{pmatrix}, \quad F = \begin{pmatrix} 0 \\ 0 \\ 0.5 \end{pmatrix} \qquad (8.10a)$$

$$C = \begin{pmatrix} 1 & 0 & 0 \end{pmatrix}, \quad J = \begin{pmatrix} 1.1 & 0 & 0 \\ 0 & 0.5 & 0 \\ 0 & 0 & 1.2 \end{pmatrix}. \qquad (8.10b)$$

The time interval considered in this example is $\Omega = [0, 5]$, the resetting times set is $\mathcal{T} = \{1.5, 3, 4.5\}$, while the IO-FTS weighting matrices are taken as $R = 0.05$ and $Q = 2$.

By simulation it can be easily checked that the system under consideration is not IO-FTS with respect to the given parameters. Figure 8.1 shows the weighted output $|y(t)|_Q^2$ of the IDLS (8.10) when the exogenous input is taken equal to 2 for all $t \in \Omega$.

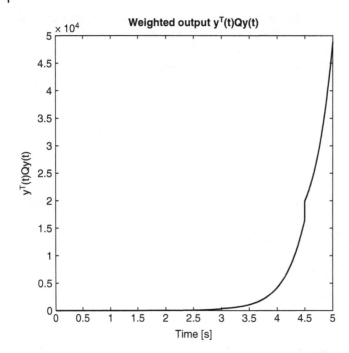

Figure 8.1 Weighted output $|y(t)|_Q^2$ of the IDLS (8.10) when the exogenous input is set equal to $w = 2$ in the time interval $\Omega = [0, 5]$.

In order to IO finite-time stabilize the considered IDLS exploiting Theorem 8.1, let us assume that the input matrix is

$$B = \begin{pmatrix} 1 & 0 & 0 \\ 0 & 1 & 0 \\ 0 & 0 & 1 \end{pmatrix}.$$

Furthermore, when solving the D/DLMI feasibility problem (8.4) by means of LMIs, the matrix-functions $T(\cdot)$, $S(\cdot)$, $\hat{A}_K(\cdot)$, $\hat{B}_K(\cdot)$, $\hat{C}_K(\cdot)$ and $D_K(\cdot)$ have been approximated by piecewise affine functions with jumps in correspondence of the resetting times, according to Appendix C.2.

Once the feasibility problem, constrained by conditions (8.4), is solved, the actual controller matrix-valued functions are computed following Procedure 8.1. Figure 8.2 shows both the three control inputs and the weighted output $|y(t)|_Q^2$ for the closed-loop system, showing the effectiveness of the proposed approach for the synthesis of the controller (8.2).

8.3 IO Finite-Time Stabilization of IDLSs: $\mathcal{W}_\infty$ Signals

We have that, for $G(\cdot) = 0$, Theorem 7.2 displays the same statement as Theorem 7.1, if we replace $Q(t)$ by $\tilde{Q}(t) := (t - t_0)Q(t)$. The following couple of results is stated here for the first time.

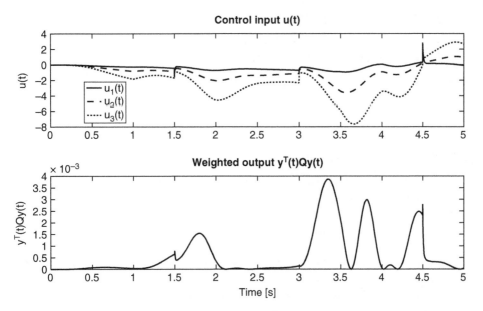

Figure 8.2 Control inputs $u(\cdot)$ and weighted output $|y(t)|_Q^2$ for the IDLS (8.10) when it is IO finite-time stabilized by a controller designed exploiting Theorem 8.1.

Theorem 8.3 (Sufficient condition for IO finite-time stabilization of IDLSs via output feedback; $\mathcal{W}_\infty$ case) Problem 8.1 is solvable, in presence of $\mathcal{W}_\infty$ signals, *if* there exist two symmetric matrix-valued functions $T(\cdot)$ and $S(\cdot)$ that are piecewise continuously differentiable in each interval $[t_k, t_{k+1})$, with $k = 0, \dots, \nu$, and $t_{\nu+1} = t_0 + T$, and four piecewise continuous matrix-valued functions $\hat{A}_K(\cdot)$, $\hat{B}_K(\cdot)$, $\hat{C}_K(\cdot)$, $D_K(\cdot)$, and 2ν matrices $\hat{J}_{K_i}$, and $\hat{H}_{K_i}$, with $i = 1, \dots, \nu$, such that the coupled D/DLMI/LMI (8.4a) and (8.4a) are satisfied, together with

$$\begin{pmatrix} \Psi_{11}(t) & \Psi_{12}(t) & 0 \\ \Psi_{12}^T(t) & S(t) & (t-t_0)^{1/2}S(t)C^T(t) \\ 0 & (t-t_0)^{1/2}C(t)S(t) & Q^{-1}(t) \end{pmatrix} > 0, \quad \forall t \in \Omega \qquad (8.11)$$

where

$$\Psi_{11}(t) = T(t) - C^T(t)\tilde{Q}(t)C(t),$$
$$\Psi_{12}(t) = I - C^T(t)\tilde{Q}(t)C(t)S(t). \qquad \blacktriangle$$

As for the state feedback problem, we have the following theorem.

Theorem 8.4 (Sufficient condition for IO finite-time stabilization of IDLSs via state feedback; $\mathcal{W}_\infty$ case) Problem 8.1, with $C(\cdot) = I$, in presence of $\mathcal{W}_\infty$ signals, is solvable via state feedback control *if* there exist a positive definite matrix-valued function $\Upsilon(\cdot)$ that is piecewise continuously differentiable in each interval $[t_k, t_{k+1})$, with $k = 0, \dots, \nu$ and $t_{\nu+1} = t_0 + T$, and a piecewise continuous matrix-valued function $L(\cdot)$, such that the coupled D/DLMI/LMI (8.9a) and (8.9b) are satisfied, together with

$$\begin{pmatrix} \Upsilon(t) & (t-t_0)^{1/2}\Upsilon(t)C^T(t) \\ (t-t_0)^{1/2}C(t)\Upsilon(t) & Q^{-1}(t) \end{pmatrix} > 0, \quad \forall t \in \Omega \qquad (8.12)$$

In this case a controller gain that solves Problem 8.1 via state feedback is

$$K(t) = L(t)\Upsilon^{-1}(t), \quad t \in \Omega.$$

▲

8.3.1 Illustrative example: Cont'd

Let us consider again the IDLS defined by the time-varying dynamic (7.33), the resetting law (7.34), and the resetting times set (7.35), which was introduced in Section 7.4.1. If we now assume that

$$R = 1, Q = 1, \Omega = [0, 2],$$

it turns out that the IDLS under consideration is not IO-FTS wrt $(\Omega, \mathcal{W}_\infty, Q)$ for this choice of the parameters. In order to stabilize the IDLS on a finite-time horizon, we can design a state feedback control law $u(\cdot) = K(t)x(t)$, when

$$B = \begin{pmatrix} 1 \\ 1 \end{pmatrix}.$$

The state feedback control can be designed exploiting Theorem 8.4; hence, by solving the corresponding D/DLMI/LMI feasibility problem. In order to solve conditions (8.9a), (8.9b), and (8.12), with $\tilde{Q} = t \cdot Q$, similarly to what has been done in Section 8.2.1, we

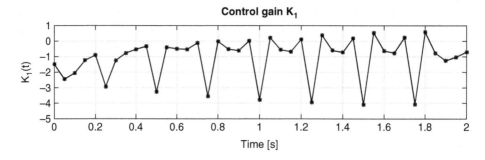

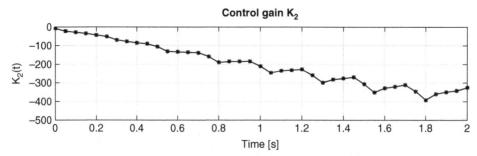

Figure 8.3 State-feedback controller gains obtained by solving the feasibility problem (8.9) for IDLS considered in Section 8.3.1.

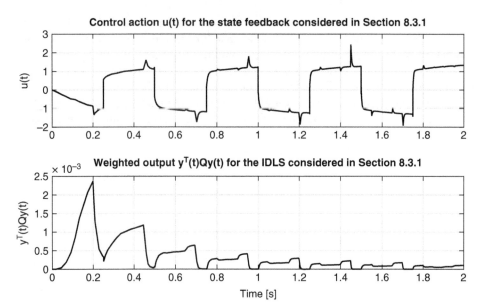

Figure 8.4 Control action $u(\cdot)$ and weighted output $|y(t)|_Q^2$ for the closed-loop system considered in Section 8.3.1.

assume a piecewise affine structure for the optimization matrix-valued functions $\Upsilon(\cdot)$ and $L(\cdot)$ with jumps in correspondence of the resetting times (7.35).

With this choice for the optimization functions, and by setting the sampling time used for their discretization equal to $T_s = 0.05$, it turns out that the feasibility problem constrained by the coupled D/DLMI/LMI (8.9) admits a solution. In particular, Figure 8.3 shows the time behavior of the two components of the state feedback control gain

$$K(t) = L(t)\Upsilon^{-1}(t) = \begin{pmatrix} K_1(t) \\ K_2(t) \end{pmatrix}.$$

Furthermore, Figure 8.4 shows both the control action $u(t) = K(t)x(t)$ and the weighted output $|y(t)|_Q^2$ for the closed-loop system.

8.4 Summary

In this chapter, the problem of the IO finite-time stabilization of IDLSs, both in presence of $\mathcal{W}_2$ and $\mathcal{W}_\infty$ exogenous inputs, has been investigated. The starting point is given by the analysis results provided in Chapter 7.

If the system state is not available for feedback, starting from condition iii) in Theorem 7.1, a necessary and sufficient condition for IO finite-time stabilization has been obtained when the system is subject to $\mathcal{W}_2$ exogenous inputs; to this end, the nonlinear change of matrix variable proposed in [98] has been again exploited.

By following essentially the same machinery, a sufficient condition for stabilization when $\mathcal{W}_\infty$ exogenous inputs are considered has been derived, starting from Theorem 7.2.

The theoretical part of the chapter is completed by a couple of conditions for the existence of a state feedback controller that IO finite-time stabilizes the closed-loop IDLS; such conditions, as usual, are obtained through the change of matrix variable $K(t) = L(t)\Upsilon^{-1}(t)$, due to Geromel and co-workers [100], where $K(\cdot)$ is the controller gain, and $\Upsilon(\cdot)$ is the matrix function associated to the Lyapunov function.

The technique developed to face the IO finite-time stabilization problem for IDLSs, both in the $\mathcal{W}_2$ and $\mathcal{W}_\infty$ cases, has been illustrated through some numerical examples.

9

Impulsive Dynamical Linear Systems with Uncertain Resetting Times

In this chapter we consider the case where the resetting times of the IDLS under consideration are uncertain. In particular, sufficient conditions that guarantee that a given IDLS, under various assumptions on the resetting times knowledge, is IO-FTS over a specified time interval are provided, for both the $\mathcal{W}_2$ and $\mathcal{W}_\infty$ input classes.

First, We consider the case of arbitrary switching, i.e., no knowledge about the resetting times is available; then, we investigate the more realistic case where the resetting times are known within a given uncertainty.

The proposed sufficient conditions for IO-FTS lead to coupled DLMI/LMIs feasibility optimization problems and therefore can be easily framed into the LMIs framework, by following the approach of Appendix C.

The methodology is illustrated by means of a numerical example in Section 9.3, which shows how the the procedure becomes more conservative as the level of uncertainty increases.

In this chapter, for the sake of brevity, only the analysis conditions are derived, since the design can be easily accomplished by following the same guidelines of Sections 8.2 and 8.3.

The content of this chapter is mostly based on the results presented in [124].

9.1 Arbitrary Switching

The case of no knowledge of the resetting times, namely arbitrary switching (AS), is tackled in this section. The main difference between the AS case and the certain case, presented in Chapter 7, is that the optimization matrix $P(\cdot)$ cannot exhibit any jumps in the interval Ω, since the resetting times are unknown.

The next theorem provides sufficient conditions for the IO-FTS of the IDLS (1.14) wrt $\mathcal{W}_2$ inputs in the AS case.

Theorem 9.1 (Sufficient condition for IO-FTS in the AS case, $\mathcal{W}_2$ inputs) Given the time interval Ω, the class of inputs $\mathcal{W}_2$, and the continuous, positive definite matrix-valued function $Q(\cdot)$, system (1.14), with $G(\cdot) = 0$, is IO finite-time stable

Finite-Time Stability: An Input-Output Approach, First Edition.
Francesco Amato, Gianmaria De Tommasi, and Alfredo Pironti.
© 2018 John Wiley & Sons Ltd. Published 2018 by John Wiley & Sons Ltd.

wrt $(\Omega, \mathcal{W}_2, Q(\cdot))$ under AS, if the coupled DLMI/LMIs

$$\begin{pmatrix} \dot{P}(t) + A(t)^T P(t) + P(t)A(t) & P(t)F(t) \\ F^T(t)P(t) & -R \end{pmatrix} < 0 \qquad (9.1a)$$

$$J^T(t)P(t)J(t) - P(t) < 0 \qquad (9.1b)$$

$$P(t) > C^T(t)Q(t)C(t), \qquad (9.1c)$$

admit a piecewise continuously differentiable, positive definite solution $P(\cdot)$, for $t \in \Omega$. ▲

Proof: Let t_i, $i = 1, \ldots, v$, be an arbitrary sequence of resetting times.

In the interval $[0, t_1)$ conditions (7.10a) and (7.10c) in Theorem 21 are satisfied in view of (9.1a) and (9.1c).

Due to the continuity of $P(\cdot)$, the satisfaction of condition (7.10b), for $i = 1$, is guaranteed from (9.1b). For $t \in (t_1, t_2)$, again conditions (7.10a) and (7.10c) are satisfied by (9.1a) and (9.1c).

Iterating these arguments, the whole interval Ω is covered; from this, the proof follows. ◇

Remark 9.1   Note that, differently from Theorem 7.1, condition (9.1b) has to be guaranteed in the whole time interval; this unavoidably leads to more conservativeness. In particular, since $P(\cdot)$ is continuous, inequality (9.1b) implies that the matrix-valued function $J(\cdot)$ has to be Schur for all $t \in \Omega$. Hence, due to the lack in the resetting times knowledge, it is necessary to have *stable* resetting laws in order to meet conditions (9.1).   ◇

Following the same considerations done for the proof of Theorem 9.1 and exploiting the statement of Theorem 7.2, a similar result can be obtained for the case of $\mathcal{W}_\infty$ signals.

Theorem 9.2 (Sufficient condition for IO-FTS in the AS case, $\mathcal{W}_\infty$ inputs)   Given the time interval Ω, the class of signals $\mathcal{W}_\infty$, and the continuous, positive definite matrix-valued function $Q(\cdot)$, system (1.14) is IO finite-time stable wrt $(\Omega, \mathcal{W}_\infty, Q(\cdot))$ under AS, if there exists a piecewise continuous scalar function $\theta(\cdot)$, with $\theta(t) > 1$ and $t \in \Omega$, such the coupled DLMI/LMIs

$$\begin{pmatrix} \dot{P}(t) + A^T(t)P(t) + P(t)A(t) & P(t)F(t) \\ F^T(t)P(t) & -R \end{pmatrix} < 0 \qquad (9.2a)$$

$$J^T(t)P(t)J(t) - P(t) < 0 \qquad (9.2b)$$

$$\theta(t)R(t) - R(t) > 2\theta(t)G^T(t)Q(t)G(t) \qquad (9.2c)$$

$$P(t) > 2\theta(t)C^T(t)\tilde{Q}(t)C(t), \qquad (9.2d)$$

where $\tilde{Q}(t) = (t - t_0)Q(t)$, admit a piecewise continuously differentiable, positive definite solution $P(\cdot)$, for $t \in \Omega$. ▲

9.2   Uncertain Switching

Let us now consider the IDLS (1.14) with uncertain switching (US), i.e., the case where the j-th resetting time is known with a given uncertainty $\pm\Delta T_j$.

Even in the US case, the sufficient condition to be checked to assess IO-FTS turns out to be more conservative with respect to the one derived in Chapter 7. Furthermore, a trade-off between uncertainty on the resetting times and additional constraints, to be added in order to check IO-FTS, is needed. In particular, the less uncertainty on the resetting times, the fewer additional constraints to be verified.

In the US case, it is useful to introduce the following definitions to describe the uncertainty on the resetting times.

$$\psi_1 =]t_0, t_1 + \Delta T_1[,$$
$$\psi_j =]t_{j-1} - \Delta T_{j-1}, t_j + \Delta T_j[, \; j = 2, \dots, v$$
$$\psi_{v+1} =]t_v - \Delta T_v, t_0 + T[$$
$$\phi_j = [t_j - \Delta T_j, t_j + \Delta T_j], \; j = 1, \dots, v$$

Furthermore, in the following we assume that

$$\bigcap_{j=1}^{v} \phi_j = \emptyset, \tag{9.3}$$

which implies the knowledge of the resetting times order.

Theorem 9.3 (Sufficient condition for IO-FTS in the US case, W_2 inputs) Given the time interval Ω, the class of inputs W_2, and the continuous, positive definite matrix-valued function $Q(\cdot)$, system (1.14), with $G(\cdot) = 0$, is IO finite-time stable wrt $(\Omega, W_2, Q(\cdot))$ under US, if there exist $v + 1$ piecewise continuously differentiable, positive definite matrix-valued functions P_j, with $j = 1, \dots, v + 1$, that satisfy the set of coupled DLMI/LMIs

$$\begin{pmatrix} \dot{P}_j(t) + A^T(t)P_j(t) + P_j(t)A(t) & P_j(t)F(t) \\ G^T(t)P_j(t) & -R \end{pmatrix} < 0$$
$$t \in \psi_j, \; j = 1, \dots, v + 1 \tag{9.4a}$$
$$J^T(t)P_{j+1}(t)J(t) - P_j(t) < 0, \quad t \in \phi_j, \; j = 1, \dots, v \tag{9.4b}$$
$$P_j(t) > C^T(t)Q(t)C(t), \quad t \in \psi_j, \; j = 1, \dots, v + 1. \tag{9.4c}$$

▲

Proof: The proof readily follows from that one of Theorem 7.1, by exploiting the knowledge of the resetting times order implied by (9.3). Indeed, condition (9.3) allows us to *assign* a single optimization matrix $P_j(\cdot)$ to each time interval ψ_j, with $j = 1, \dots, v + 1$.

Notice that condition (9.4b) has to be checked in ϕ_j, i.e., the time interval in which the state jump could occur. ◇

It should be noted that the length of ψ_j and ϕ_j decrease when the uncertainties get smaller, leading us to the same result of Theorem 7.1 when $\Delta T_j = 0$ for all j.

The following result is a sufficient condition for IO-FTS in presence of W_∞ signals. The proof can be derived following the same arguments as in the one of Theorem 9.3.

Theorem 9.4 (Sufficient condition for IO-FTS in the US case, W_∞ inputs) Given the time interval Ω, the class of inputs W_∞, and the continuous, positive definite matrix-valued function $Q(\cdot)$, system (1.14) is IO finite-time stable wrt $(\Omega, W_\infty, Q(\cdot))$

under US, if there exist $v + 1$ piecewise continuously differentiable, positive definite matrix-valued functions P_j, $j = 1, \ldots, v + 1$, and $v + 1$ piecewise continuous scalar functions $\theta_j(\cdot)$, with $\theta_j(t) > 1, j = 1, \ldots, v + 1$, and $t \in \Omega$, that satisfy the set of coupled DLMI/LMIs

$$\begin{pmatrix} \dot{P}_j(t) + A^T(t)P_j(t) + P_j(t)A(t) & P_j(t)F(t) \\ G^T(t)P_j(t) & -R \end{pmatrix} < 0$$

$$t \in \psi_j, \quad j = 1, \ldots, v + 1 \tag{9.5a}$$

$$J^T(t)P_{j+1}(t)J(t) - P_j(t) < 0, \quad t \in \phi_j, \quad j = 1, \ldots, v \tag{9.5b}$$

$$\theta_j(t)R(t) - R(t) > 2\theta_j(t)G^T(t)Q(t)G(t), \quad t \in \psi_j, \quad j = 1, \ldots, v + 1 \tag{9.5c}$$

$$P_j(t) > 2\theta_j(t)C^T(t)\tilde{Q}(t)C(t), \quad t \in \psi_j, \quad j = 1, \ldots, v + 1, \tag{9.5d}$$

where $\tilde{Q}(t) := (t - t_0)Q(t)$.   ▲

9.3 Numerical Example

This section presents a numerical example to show the effectiveness of the proposed approach, when checking the IO–FTS of IDLSs in the case of different levels of knowledge on the resetting times. In particular, we focus our attention on input signals belonging to the class W_∞.

Let us consider the second-order IDLS that is equivalent to the switching linear system (SLS, see Section 1.4.2) defined by the following two linear systems

$$S_1 : \begin{cases} \dot{x}(t) = \begin{pmatrix} 0.5 + 0.5t & 0.1 \\ 0.4 & -0.3 + 0.5t \end{pmatrix} x(t) + \begin{pmatrix} 1 \\ 1 \end{pmatrix} w(t) \\ y(t) = \begin{pmatrix} 1 & 1 \end{pmatrix} x(t), \end{cases} \tag{9.6}$$

$$S_2 : \begin{cases} \dot{x}(t) = \begin{pmatrix} 0.15 + 0.5t & 0.2 \\ 1 & -0.25 - 0.5t \end{pmatrix} x(t) + \begin{pmatrix} 1 \\ 0 \end{pmatrix} w(t) \\ y(t) = \begin{pmatrix} 2 & 1 \end{pmatrix} x(t), \end{cases} \tag{9.7}$$

and by the switching signal $\sigma(\cdot)$, whose time trace is reported in Figure 9.1 for the time interval $\Omega = [0, 1]$. Such switching signal induces the following resetting times set

$$\mathcal{T} = \{0.2, 0.5, 0.75\}.$$

Moreover, the considered resetting law is given by

$$J = \begin{pmatrix} -1.1 & 0 \\ 0 & 0.1 \end{pmatrix}. \tag{9.8}$$

Hence, the IDLS equivalent to the considered SLS, will *run* the dynamic S_2 in the time intervals $[0, 0.2]$ and $[0.5, 0.75]$, while it will *run* the dynamic S_1 in the time intervals $[0.2, 0.5]$ and $[0.75, 1]$. In correspondence of each resetting time, the resetting law (9.8) will apply.

For what concerns the IO-FTS, we consider the following parameters

$$R = 1, \quad \Omega = [0, 1]. \tag{9.9}$$

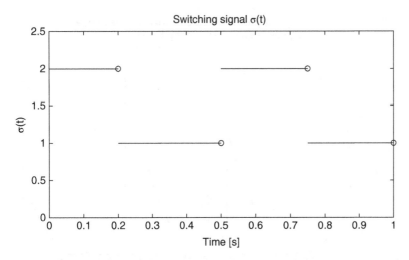

Figure 9.1 Switching signal $\sigma(\cdot)$ for SLS considered in Section 9.3.

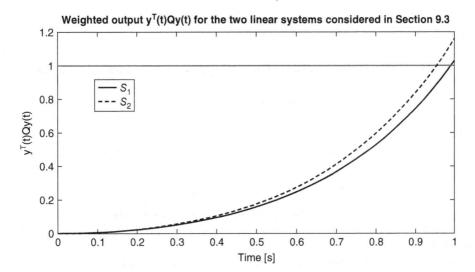

Figure 9.2 Weighted output $|y(t)|_Q^2$ for the two linear systems considered in Section 9.3, when $w(t) = 1$ in Ω, and $Q = 0.12$.

Before considering the different cases introduced in Section 9.1 and 9.2, note that both linear systems considered in this example are unstable. Furthermore, they are also not IO finite-time stable when considering the parameters specified in (9.9), and when $Q \geq 0.12$, as it is shown in Figure 9.2, where the weighted output for $w(t) = 1$ in Ω is reported.

9.3.1 Known resetting times

If there is no uncertainty on the resetting times, the feasibility problem associated to Theorem 7.2 can be recast into LMIs by means of a piecewise affine matrix-valued

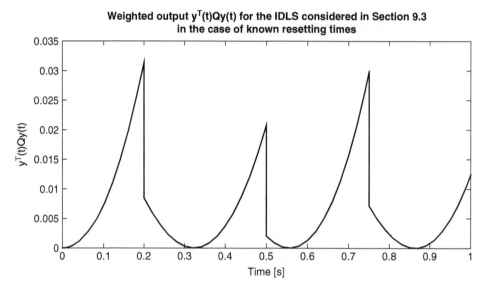

Figure 9.3 Weighted output $|y(t)|_Q^2$ for the IDLS considered in Section 9.3, when $w(t) = 1$ in Ω, and $Q = 0.17$.

function $P(\cdot)$ with jumps in correspondence of the resetting times, according to Appendix C.2.

Given a value of Q, by using an LMI solver such as the the Matlab LMI toolbox ([90]), it is possible to solve the feasibility problem (7.28); in particular, Figure 9.3 shows the weighted output $|y(t)|_Q^2$, for $Q = 0.17$, when $w(t) = 1$, in the interval Ω. It follows that this value of Q makes the IDLS IO finite-time stable in the interval $\Omega = [0, 1]$.

9.3.2 Arbitrary switching

If we now assume that $\mathcal{T}$ is totally unknown, in order to recast the D/DLMI condition of Section 9.1 in terms of LMIs, the matrix-valued function $P(\cdot)$ must be assumed piecewise affine without jumps in the time interval Ω.

However, as it was remarked in Section 9.1, since the J matrix is not Schur, the feasibility problem (9.2) does not admit any solution, no matter how we choose the weighting matrix $Q > 0$ (see also Remark 9.1).

9.3.3 Uncertain switching

Let us now consider the case of uncertain switching with $\Delta T_j = 0.03$, when $j = 1, \dots, 3$. Even in this case, the D/DLMIs feasibility problem (9.5) can be recast into LMIs by choosing the matrix variables $P_i(\cdot)$, $i = 1, \dots, 4$, with a piecewise affine structure in the interval Ω.

By exploiting the MATLAB LMI Toolbox®, it turns out that the considered system is IO-FTS wrt (Ω, W_∞, Q), for all $Q \le 0.14$. Figure 9.4 shows the worst case for the weighted output, when $Q = 0.14$ and the exogenous input is equal to $w(t) = 1$ in the time interval Ω.

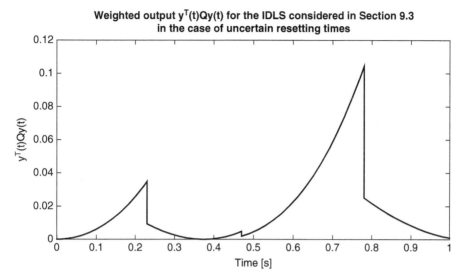

Figure 9.4 Worst case weighted output $|y(t)|_Q^2$ for the IDLS considered in Section 9.3, in the case of uncertain switching, when $Q = 0.14$, and the exogenous input is taken equal to 1 in the time interval Ω.

It turns out that, even with a partial knowledge of the switching times, it is possible to assess IO-FTS of the system, although the resetting law matrix J is not Schur.

9.4 Summary

In practical engineering problems, the occurrence of the resetting times can be unknown or, at least, uncertain; therefore, in this chapter, the IO finite-time stabilization problem for IDLSs in presence of uncertain, or totally unknown, resetting times has been considered.

The various situations have been compared with the certain case, discussed in previous chapters; as expected, the more uncertain is the knowledge of the occurrence of the resetting times, the more difficult is the satisfaction of the IO-FTS property for the system under consideration.

The first result of the chapter is a sufficient condition for IO-FTS of IDLSs in presence of arbitrary switching, i.e., the time of the occurrences of the resetting times is totally unknown, both for the $\mathcal{W}_2$ and $\mathcal{W}_\infty$ cases. In this situation, a major constraint is that we have to pick a continuous matrix-valued function $P(\cdot)$ over the whole interval Ω (remember that in the certain switching case, the Lyapunov function can switch at each resetting time, greatly relaxing both the analysis and the design problems).

Then the case of uncertain switching has been dealt with. This circumstance can be considered intermediate between the certain and uncertain switching case. From the statements of Theorems 9.3 and 9.4, it clearly appears that the degree of conservativeness becomes more severe as the level of uncertainty increases.

The chapter is ended by a numerical example that clearly illustrates the considerations above. It is worth noting that the system treated in the example is an SLS, rather than a classical IDLS (see the comments at the end of Section 1.4.2).

10

Hybrid Architecture for Deployment of Finite-Time Control Systems

In this chapter we propose a hybrid architecture that can be used to implement controllers whose design is based on finite-time stability techniques, such as the IO-FTS methodology.

As it has been shown in the previous chapters, the approach illustrated in this book to design the controller leads to a time-varying control law that cannot be applied over an infinite time interval.

To overcome this limitation, in this chapter we shall describe an architecture that makes use of a hybrid automaton [125], that combines a time-invariant controller (i.e., the *nominal* controller) with the time-varying one, which is designed exploiting the concept of IO-FTS and which is enabled when an assigned performance index exceeds a given threshold, indicating that the control performance itself is going below the desired target.

10.1 Controller Architecture

A simplified block diagram of the proposed architecture for the deployment of real-world controllers, whose design is partly based on the finite-time control techniques described in this book, is reported in Figure 10.1.

For the sake of simplicity, in this chapter we consider the case where two different continuous-time control laws are involved; they can be enabled and disabled by a hybrid automaton that acts as *supervisor*. However, the proposed architecture can be easily extended to the case of more than two controllers, as it will be briefly discussed at the end of this section.

The two continuous-time controllers in Figure 10.1

- the *steady state, LTI controller*, designed by using one among the many approaches that lead to a time-invariant control law (optimal control, $\mathcal{H}_\infty$ control, among the others);
- the *finite-time, LTV controller*, whose design is assumed to be carried out using one of the synthesis approaches presented in the previous chapters for closed-loop IO finite-time stabilization (either via state feedback or output feedback).

As said above, these two controllers are enabled by the supervisor block in Figure 10.1, according to the following switching logic. By default, the LTI controller is enabled; as soon as a given *performance index* goes below the desired minimum threshold, the

Finite-Time Stability: An Input-Output Approach, First Edition.
Francesco Amato, Gianmaria De Tommasi, and Alfredo Pironti.
© 2018 John Wiley & Sons Ltd. Published 2018 by John Wiley & Sons Ltd.

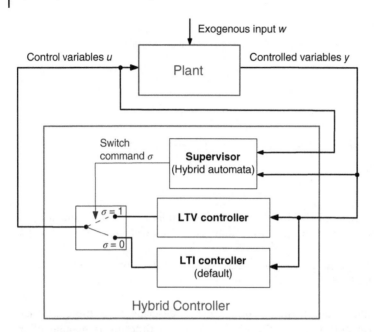

Figure 10.1 Block diagram of the proposed hybrid architecture for the implementation of a controller based on an IO-FTS design approach.

supervisor allows the switch from the LTI controller to the LTV controller. The latter is enabled for a time interval equal to the finite interval Ω used to design the IO finite-time stabilization control law. Possible choices for the performance index (some examples will be also given in Section 10.2) are:

- a functional depending on the control error;
- a functional depending on both the control and the controlled variables.

In order to implement the switching logic described above, the supervisor implements the hybrid automaton depicted in Figure 10.2, which has two discrete states and one continuous state $x_a(\cdot)$. The discrete states are:

- $S_{default}$; when this state is active, the constrained default LTI controller is enabled (the automaton output σ is set equal to 0). Given the performance index $J(\cdot)$, the guard condition (i.e., the *exit event*) becomes active when $J(\cdot) \notin \wp$, where $\wp$ is the set of

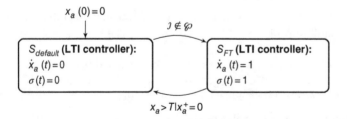

Figure 10.2 Hybrid automaton implementing the supervisor block provided in the architecture reported in Figure 10.1.

admissible values for the performance index. The exit event causes the switch to the state S_{FT}. When the state $S_{default}$ is active, the continuous state x_a is kept constant, i.e., $\dot{x}_a(t) = 0$. Note that the state $S_{default}$ is the initial state of the supervisor and that the initial value for the continuous state x_a is 0.

- S_{FT}; when this state is active, the LTV controller is enabled for T seconds, by setting the automaton output σ equal to 1 and

$$\dot{x}_a(t) = 1 \Rightarrow x_a(t) = t - \tau,$$

where τ is the time instant corresponding to the last switch $S_{default} \rightarrow S_{FT}$. This implies that T seconds after S_{FT} has been enabled, the exit event occurs and the hybrid automaton switches back to $S_{default}$.

Before introducing the two applications in the next section, it is worth remarking that:

- the number of default controllers provided in the proposed architecture can be easily increased by adding the corresponding discrete states and guard conditions in the supervisor;
- the default controller is not restricted to be linear; therefore, the proposed architecture can be adopted to deploy also time-varying nonlinear control laws;
- the finite-time controller provided in the hybrid architecture can be designed using one of the results proposed in this book, as well as other finite-time stability approaches, such as the one for *state* FTS [35];
- differently from the finite-time horizon model predictive control, where the time-varying control action is updated on line at each time sample, the LTV finite-time stabilization control law is computed off line, and once triggered, it needs to be applied for the time interval used during the design phase.

Last but not least, it should be noticed that, dealing with a hybrid controller, it is necessary to check the stability of the overall system when the supervisor switches from the LTI to the LTV controller and vice versa. This can be done by defining a common Lyapunov function for the two closed-loop systems in $S_{default}$ and S_{FT}, respectively (an example is given in Section 10.2.2, more details about stability of hybrid systems under arbitrary switching can be found in [126, Chapter 2]).

10.2 Examples

In the following we show how the hybrid architecture presented in Section 10.1 can be used to deploy an IO finite-time stabilization controller for two possible applications in the automotive field, namely the control of an active suspension and a vehicle lateral collision avoidance system.

10.2.1 Hybrid active suspension control

In Section 5.4, the problem of controlling an active suspension system has been framed within the *structured IO-FTS* context. The structured IO finite-time stabilizing state-feedback controller proposed in Section 5.4, showed better performance in rejecting the disturbances related to ground asperity, when compared with an LTI

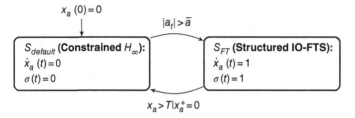

Figure 10.3 Hybrid automaton for the possible implementation of the active suspension control system based on structured IO finite-time stabilization.

controller proposed in [102], which is designed using a constrained $\mathcal{H}_\infty$ approach (see Figure 5.2)

We now propose to use the hybrid architecture described in Section 10.1 as a possible implementation of the structured IO-FTS control algorithm.

In this case, the hybrid automaton implemented by the supervisor is shown in Figure 10.3, where the default controller enabled in $S_{default}$ is the LTI controller proposed in [102].

For this example, the chosen *performance* index is taken equal to the absolute value of the tire acceleration a_t, which should be kept below the threshold $\bar{a} > 0$, set equal to 3 m/s^2. Hence, we have

$$j(t) = |a_t(t)|,$$

and

$$\wp = \{a_t \in \mathbb{R} \text{ such that } |a_t| \leq \bar{a}\}.$$

When the prescribed threshold for $j(\cdot)$ is violated, the automaton switches to the S_{FT}, enabling the structured IO finite-time state feedback designed in Section 5.4. This state remains active for $T = 2$ s, according to the finite-time interval used to design the structured IO finite-time stabilizing state feedback.

To prove the effectiveness of the proposed hybrid approach, instead of considering the isolated bump (5.20), the ground asperity shown in Figure 10.4 is considered in this section.

The simulation results are shown in Figure 10.5, where the behavior of the hybrid controller is compared with that of the constrained $\mathcal{H}_\infty$ controller. Similarly to what has been shown in Section 5.4, also in this case the proposed controller, which relies on the structured IO finite-time stabilizing state feedback, achieves a reduction of the peaks of the body acceleration. Note that the first peak is not reduced; this is due to the fact that the tire acceleration is below the threshold, and hence the state-feedback controller, designed vie the finite-time control approach, is not active.

10.2.2 Lateral collision avoidance system

In this section we exploit again the structured IO-FTS and the proposed hybrid architecture to design a vehicle lateral collision avoidance system.

In order to recast the lateral collision avoidance problem in the structured IO-FTS framework, we consider the partition of the output in Section (E.10) induced by the

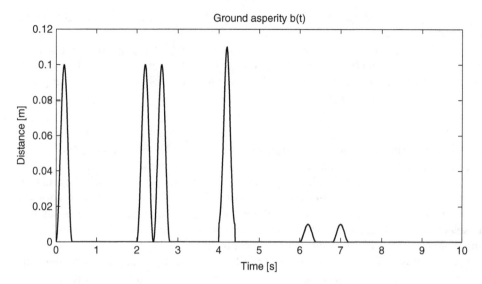

Figure 10.4 Ground asperity considered to prove the effectiveness of the hybrid controller for the active suspension system.

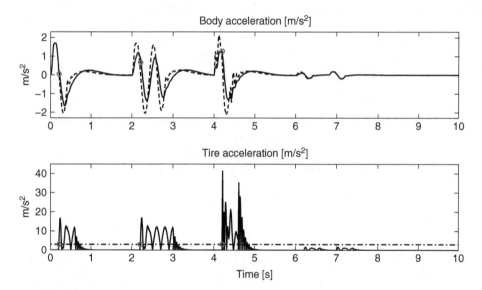

Figure 10.5 Response to the ground asperity reported in Figure 10.4. Hybrid controller (–), constrained $\mathcal{H}_\infty$ controller (- -). The circles denote the time instants when the discrete state S_{ST} of the hybrid automaton is activated.

triplet $(1, 1, 1)$, i.e. we set $\alpha = 3$ (see Definition 5.1), obtaining

$$y_1 = \dot{y}, \quad y_2 = \omega, \quad y_3 = Y.$$

By this choice, we can use the structured IO-FTS approach to limit the lateral velocity $\dot{y}$, the yaw rate ω, and the lateral displacement Y due to a side wind gust w of a given maximum amplitude and duration, by accordingly choosing the output weighting functions Q_1, Q_2, and Q_3. For the sake of simplicity, we choose these weights to be constant[1].

In order to minimize the vehicle lateral displacement due to the considered reference wind gust, we maximize the value of Q_3 subject to the constraints of Theorem 5.4. Hence the LTV structured IO finite-time stabilizing state-feedback control is computed by solving the semidefinite programming problem

$$\max_{s.t.\,(5.14a)\text{ and }(5.18)} Q_3, \tag{10.1}$$

i.e., by finding the solution to an optimization problem with DLMI constraints.

By maximizing Q_3, the lateral displacement Y_{max} is *minimized* for the class of signals in $\mathcal{W}_2$, in the time interval Ω. Indeed, by limiting Y, we achieve lateral collision avoidance in the presence of side wind when the scenario of multiple vehicles that are automatically driven on a multi-lane road is considered. By choosing the time interval Ω, it is possible to select the time horizon for which the effect of the wind gust should be minimized by the structured IO-FTS LTV state-feedback controller.

Structured IO-FTS allows us also to limit the steering angle requested by the controller, i.e., to mitigate the control action, by choosing the weighing function V_1.

The default LTI controller provided in the proposed hybrid architecture can be designed applying again a constrained $\mathcal{H}_\infty$ control approach, while the automaton implemented within the supervisor is similar to the one reported in Figure 10.3.

For the vehicle lateral collision avoidance, the *performance* index is chosen equal to the lateral displacement, i.e., $J(t) = Y(t)$. Similarly to what has been done for the active suspension considered in Section 10.2.1, the performance index $J = Y$ should remain below a given threshold $\overline{Y} > 0$, that is,

$$\wp = \{Y \in \mathbb{R} \text{ such that } |Y| \leq \overline{Y}\}.$$

When J exceeds the threshold, i.e., $J > \overline{Y}$, then the supervisor enables the structured IO-FTS controller for a time interval of length T.

Let us now briefly describe the design of the structured IO finite-time stabilizing state-feedback controller. The output weights are chosen as follows

$$Q_1 = 1/\dot{y}_{max}^2 \cong 0.51, \quad Q_2 = 1/\omega_{max}^2 \cong 268,$$

while the weight on the control action $u = \delta$ is set equal to

$$V_1 = 1/\delta_{max}^2 \cong 3283.$$

The finite-time interval is set equal to $\Omega = [0, 6]$, which is assumed to be the maximum duration of a *sustained* wind gust to be rejected. Moreover, we have set the weight R constant and equal to 1, in order to consider signals with energy less or equal to 1 in the interval Ω.

Once all the required parameters have been defined, the structured IO-FTS controller is obtained by solving the semidefinite programming problem (10.1). The optimal value for the weight on the vehicle lateral displacement is equal to $Q_3^* \cong 560$, which implies a maximum vehicle lateral displacement of about 4 cm for the considered constraints and wind energy.

---

1 Note that, given the choice of the output partition α, the output weighting functions Q_i turn out to be scalars.

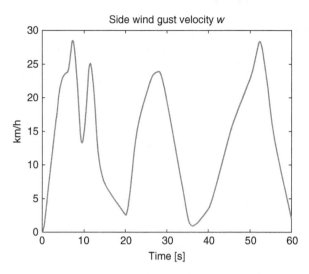

Figure 10.6 Side wind velocity profile considered as exogenous input for the simulation described in Section 10.2.2.

In order to show the effectiveness of the proposed solution for lateral collision avoidance, let us consider the velocity profile of the lateral wind gust w shown in Figure 10.6. The simulation results are summarized in Figure 10.7 when the threshold on the vehicle lateral displacement is set equal to $\overline{Y} = 10$ cm. In particular, the vehicle lateral displacement Y, and the control action δ, are shown both for the hybrid controller and in the case where the constrained $\mathcal{H}_\infty$ is used to reject the exogenous disturbance.

Note that, in order to compare the default controller with the LTV controller, the design of the LTI (default) controller has been carried out by using a constrained $\mathcal{H}_\infty$ approach, considering exogenous disturbances with bounded energy ≤ 1.

From Figure 10.7, it turns out that the structured IO-FTS methodology permits to improve the performance of the collision avoidance system, while keeping all the constrained variables within the prescribed bounds.

As it has been anticipated in Section 10.1, since the proposed lateral collision avoidance system is a hybrid system, it is necessary to check the stability of the overall system when the supervisor switches from $S_{default}$ to S_{ST} and vice versa, under an arbitrary switching sequence, i.e., under any possible pattern for the signal $\sigma(t)$.

This can be done by defining a common Lyapunov function for the two closed-loop systems when the supervisor is in the state $S_{default}$ and S_{ST}, respectively; (more details about stability of hybrid systems under arbitrary switching can be found in [126, Chapter 2]).

In order to do that, we consider the two state transitions of the hybrid automaton shown in Figure 10.2:

- $S_{default} \rightarrow S_{ST}$; in this case the state-feedback matrix switches from the value K_∞, obtained applying the constrained $\mathcal{H}_\infty$ approach, to the first value of the LTV structured IO finite-time stabilizing state-feedback controller, i.e., to $K(0)$;
- $S_{ST} \rightarrow S_{default}$; here the control matrix switches from the last value of the LTV state-feedback controller, namely $K(T)$, to K_∞.

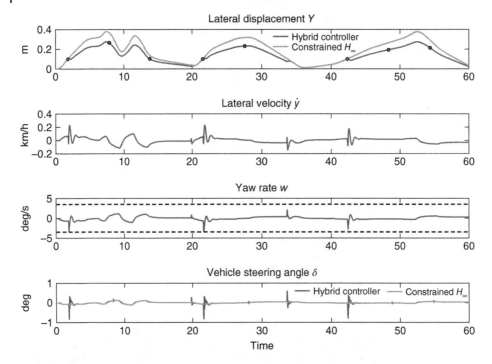

Figure 10.7 Behavior of the hybrid control architecture for vehicle collision avoidance. The circles denote the time instants when the state S_{ST} of the hybrid automaton is activated, that is, the time instants when the event $Y > \overline{Y}$ occurs, causing the activation of the IO finite-time stabilizing controller for a time window of length T.

It follows that, a sufficient condition that assures internal stability of the hybrid vehicle lateral collision avoidance system under any switching sequence is given by the following theorem, which follows from [126, Theorem 2.1]. The sufficient condition involves the solution of an LMI feasibility problem.

Theorem 10.1 Consider the two-wheel vehicle (E.10) and the hybrid control system described in Section 10.1. If there exist a positive definite matrix P such that the following inequalities are satisfied

$$(A + BK_\infty)^T P + P(A + BK_\infty) < 0 \qquad (10.2a)$$

$$(A + BK(0))^T P + P(A + BK(0)) < 0 \qquad (10.2b)$$

$$(A + BK(T))^T P + P(A + BK(T)) < 0, \qquad (10.2c)$$

then the closed-loop system is internally stable under any arbitrary switching sequence between the two states $S_{default}$ and S_{ST} of the supervisor shown in Figure 10.2 ▲

Let us consider the two-wheel model, the time interval $\Omega = [0, 6]$, and the following values for the state-feedback control matrices

$$K_\infty = (-0.162 \ -0.895 \ -9.794 \ -1.394)$$

$$K(0) = (-0.159 \ -0.255 \ -3.700 - 0.722) \cdot 10^5,$$

$$K(6) = (-0.159 - 0.256 - 3.714 - 0.725) \cdot 10^5.$$

It turns out that the LMI feasibility problem (10.2) admits the following solution

$$P = \begin{pmatrix} 0.0008 & -0.0010 & 0.0089 & -0.0005 \\ -0.0010 & 0.0014 & -0.0112 & 0.0008 \\ 0.0089 & -0.0112 & 0.2413 & 0.0032 \\ -0.0005 & 0.0008 & 0.0032 & 0.0153 \end{pmatrix},$$

hence, the overall hybrid system is stable.

10.3 Summary

As stated before, the IO-FTS approach is useful to refine the system behavior during the transient phase, while classical IO (Lyapunov) stability is a fundamental requirement to guarantee the correct behavior at steady state; therefore, it is good practice to satisfy both requirements when designing a control system.

To this end, in this chapter we have discussed an architecture that makes use of an hybrid automaton, which combines the nominal time-invariant controller with a time-varying one, designed exploiting the concept of IO-FTS, and which is enabled when a given performance index exceeds an assigned threshold, indicating that the control performance itself is going below the desired target.

This approach has been illustrated through a couple of examples. Such examples are even more interesting, since they describe a practical application of the structured IO-FTS approach detailed in Chapter 5.

The first example exploits the IO finite-time stabilizing controller, designed in Section 5.4, to control an active suspension system; such controller showed better performance in rejecting the disturbances related to ground asperity when compared with an LTI controller designed using an $\mathcal{H}_\infty$ approach. In this chapter, the design has been completed by realizing the hybrid architecture described in Section 10.1, which also guarantees a stable steady-state behavior.

In the latter example, the hybrid architecture has been exploited to design a lateral collision avoidance controller for the two-wheel system described in Section E.4. The structured IO finite-time stabilizing methodology has been exploited to impose, during the transient phase, separate bounds for each of the controlled variables, namely the lateral velocity $\dot{y}$, the yaw rate ω, and the lateral displacement Y, together with the steering angle (the control action), when the system is subject to a side wind gust disturbance. The stability of the whole system (steady state $\mathcal{H}_\infty$ plus IO finite-time stabilizing controller) has been proven with the aid of a common quadratic Lyapunov function.

A

Fundamentals on Linear Time-Varying Systems

In this appendix we consider the qualitative behavior of solutions of the system of linear differential equations

$$\dot{x}(t) = A(t)x(t), \qquad t \geq 0, \tag{A.1}$$

where $x(t) \in \mathbb{R}^n$. In particular we shall investigate which hypothesis the matrix function $A(\cdot)$ must satisfy such that *existence* and *uniqueness* of the solution of system (A.1) are guaranteed. Moreover the definition of state-transition matrix is recalled, together with the main definitions involving classical Lyapunov stability.

Most of the content of this and the next section is taken from [96].

A.1   Existence and Uniqueness

Our main result of the section is a theorem guaranteeing existence and uniqueness of the solution of system (A.1). We recall that if $X(\cdot)$ is a piecewise continuous matrix-valued function, in any compact interval contained in $\mathbb{R}_0^+$ it has a finite number of discontinuity points; at a discontinuity point the left and right limits of $X(\cdot)$ exist and are finite.

Theorem A.1 (Existence and uniqueness of the solution)   Let $t_0 \geq 0$, $x_0 \in \mathbb{R}^n$ and assume that $A(\cdot)$ is piecewise continuous over $\mathbb{R}_0^+$; then system (A.1) admits a *unique* continuous solution $\varphi(\cdot, t_0, x_0)$, defined for $t \geq t_0$, satisfying $\varphi(t_0, t_0, x_0) = x_0$.   ▲

Proof: Since $A(\cdot)$ is piecewise continuous, for any given x, $A(t)x$ is a bounded and continuous function of t, with the exception of at most a finite number of points in any compact subset of $[t_0, +\infty)$. Moreover, for arbitrary $x, y \in \mathbb{R}^n$, $L > 0$,

$$
\begin{aligned}
|A(t)x - A(t)y| &\leq |A(t)| \, |x - y| \\
&\leq k_L |x - y| \\
&:= \sup_{t \in [t_0, t_0 + L]} |A(t)| \, |x - y|, \qquad t \in [t_0, t_0 + L]; \tag{A.2}
\end{aligned}
$$

that is the RHS of (A.1) satisfies a global Lipschitz condition in x. This allows to apply the existence and uniqueness theorem for differential equations (see [93], pp. 470–471).   ◊

In the following, we shall assume that the hypothesis of Theorem A.1 hold for $A(\cdot)$.

Finite-Time Stability: An Input-Output Approach, First Edition.
Francesco Amato, Gianmaria De Tommasi, and Alfredo Pironti.
© 2018 John Wiley & Sons Ltd. Published 2018 by John Wiley & Sons Ltd.

A.2   The State Transition Matrix

Let us denote by $x(t) = \varphi(t, t_0, x_0)$ the unique solution of system (A.1) starting from x_0 at time t_0. Now it is simple to show that, for a given pair $(t, t_0) \in \mathbb{R}_0^+ \times \mathbb{R}_0^+$, the mapping

$$x_0 \in \mathbb{R}^n \mapsto \varphi(t, t_0, x_0) \in \mathbb{R}^n \tag{A.3}$$

is linear. Hence by the Matrix Representation Theorem [77, p. 188], there exists a matrix $\Phi(t, t_0)$ such that

$$\varphi(t, t_0, x_0) = \Phi(t, t_0)x_0. \tag{A.4}$$

The matrix function $(t, t_0) \in \mathbb{R}_0^+ \times \mathbb{R}_0^+ \to \Phi(t, t_0)$ is called the *state-transition matrix*. It plays a fundamental role for the study of linear time-varying systems. The next result is obvious.

Theorem A.2 (Solution of system (A.1))   The unique solution of system (A.1) starting at time t_0 from x_0 is

$$x(t) = \Phi(t, t_0)x_0. \tag{A.5}$$

▲

Note that, for all $x_0 \in \mathbb{R}^n$, we have $\Phi(t_0, t_0)x_0 = x(t_0) = x_0$. From this follows that

$$\Phi(t_0, t_0) = I. \tag{A.6}$$

Equality (A.6) is referred to as the *consistency* property of the state-transition matrix.

Definition A.1 (Fundamental Matrix)   Any solution $X(\cdot)$ of the matrix differential equation

$$\dot{X}(t) = A(t)X(t), \tag{A.7}$$

satisfying $\det(X(t)) \neq 0$ for all $t \in \mathbb{R}_0^+$, is called a *Fundamental Matrix* of system (A.1). ◇

Theorem A.3 (Computation of Φ via a fundamental matrix)   The following equality holds for all $t, t_0 \in \mathbb{R}_0^+, t \geq t_0$,

$$\Phi(t, t_0) = X(t)X^{-1}(t_0) \tag{A.8}$$

where $X(\cdot)$ is any fundamental matrix of system (A.1). ▲

Proof: The proof follows from the fact that both sides of (A.8) satisfy the same matrix differential equation

$$\dot{\Lambda}(t) = A(t)\Lambda(t), \qquad \Lambda(t_0) = I. \tag{A.9}$$

◇

The next properties follow directly from Theorem A.3

Theorem A.4 (Composition (Transition))   For all $t, t_0, t_1 \in \mathbb{R}_0^+, t \geq t_1 \geq t_0$,

$$\Phi(t, t_0) = \Phi(t, t_1)\Phi(t_1, t_0). \tag{A.10}$$

▲

From Theorem A.4 it follows that, letting

$$x_1 = \varphi(t_1, t_0, x_0), \qquad t_1 > t_0,$$

we have

$$\varphi(t, t_0, x_0) = \varphi(t, t_1, x_1), \qquad t \geq t_1. \tag{A.11}$$

Equality (A.11) is called *Transition Property* of the state; this explains the name given to $\Phi(t, t_0)$.

By (A.8) we can extend the definition of $\Phi(\cdot, \cdot)$ to the case in which the first argument is not greater than the second argument, indeed for $0 \leq t_0 \leq t$

$$\Phi(t_0, t) := X(t_0)X^{-1}(t). \tag{A.12}$$

The next result, which derives directly from (A.8) and (A.12), shows that $\Phi(t_0, t)$ is exactly the inverse of $\Phi(t, t_0)$.

Theorem A.5 (Inversion) For all $(t, t_0) \in \mathbb{R}_0^+ \times \mathbb{R}_0^+$ we have

$$\Phi(t, t_0)^{-1} = \Phi(t_0, t). \tag{A.13}$$

▲

Note that, for all $(t, t_0) \in \mathbb{R}_0^+ \times \mathbb{R}_0^+$, $\Phi(t, t_0)$ is always invertible; this means that, for all $0 \leq t_0 \leq t$, it is always possible to go back in time and obtain x_0 starting from $x(t)$:

$$x_0 = \Phi(t, t_0)^{-1}x(t)$$
$$= \Phi(t_0, t)x(t). \tag{A.14}$$

Example A.1 Let us consider system (A.1) with

$$A(t) = \begin{pmatrix} -1 - 5\cos t \sin t & -5\cos^2 t + 1 \\ 5\sin^2 t - 1 & -1 + 5\cos t \sin t \end{pmatrix}. \tag{A.15}$$

It is simple to verify that, according to Theorem A.3, the state-transition matrix is given by (A.8), where

$$X(t) = e^{-t} \begin{pmatrix} \cos t & -5t \cos t + \sin t \\ -\sin t & 5t \sin t + \cos t \end{pmatrix}. \tag{A.16}$$

△

In general, the analytical computation of the transition matrix is not a simple task. The following theorem provides a way of computing (with some approximation) $\Phi(t, t_0)$ as the partial sum of the so-called *Peano-Baker* series.

Theorem A.6 (Peano-Baker series, [93], p. 13) The Peano-Baker series

$$I + \int_{t_0}^{t} A(\tau_1)d\tau_1 + \int_{t_0}^{t} A(\tau_1)\left[\int_{t_0}^{\tau_1} A(\tau_2)d\tau_2\right] d\tau_1$$

$$+ \int_{t_0}^{t} A(\tau_1)\left[\int_{t_0}^{\tau_1} A(\tau_2)\left[\int_{t_0}^{\tau_2} A(\tau_3)d\tau_3\right] d\tau_2\right] d\tau_1 + \cdots \tag{A.17}$$

uniformly converges to the state-transition matrix $\Phi(t, t_0)$ over any compact interval of $\mathbb{R}_0^+$. ▲

The next theorem deals with the important case of linear time-invariant systems.

Theorem A.7 (State-transition matrix for LTI systems, [93, p. 71]) Assume that $A(\cdot) = A \in \mathbb{R}^{n \times n}$. In this case we have

$$\Phi(t, t_0) = \Phi(t - t_0)$$

$$= \sum_{i=0}^{+\infty} \frac{A^i(t - t_0)^i}{i!}$$

$$=: \exp(A(t - t_0)).$$

▲

Proof: The proof follows from Theorem A.6 and the fact that

$$\int_{t_0}^t A(\tau_1) \left[\int_{t_0}^{\tau_1} A(\tau_2) \left[\cdots \left[\int_{t_0}^{\tau_{i-1}} A(\tau_i) d\tau_i \right] d\tau_{i-1} \right] \cdots d\tau_2 \right] d\tau_1 = \frac{A^i(t - t_0)^i}{i!}.$$

$$(A.18)$$

◇

The evaluation of $\Phi(t, t_0)$ in the time-invariant case can be performed numerically by computing the partial sum of the series $\sum_{i=0}^{+\infty} \frac{A^i(t-t_0)^i}{i!}$.

However, in the time-invariant case, the transition matrix can be also evaluated in closed form either via the Laplace transform method or by performing a similarity transformation on A, to put it in Jordan form (see Chaps. 3 and 4 in [93]).

A.3 Lyapunov Stability of Linear Time-Varying Systems

In the following the various definitions of Lyapunov stability of the equilibrium point $x = 0$ of system (A.1) are recalled; remember that $\varphi(\cdot, t_0, x_0)$ is the solution starting from x_0 at time t_0.

Good sources for Lyapunov stability theory are the books [127], [128], [32], [129] and the paper by Kalman [130].

Definition A.2 (Lyapunov stability) The equilibrium point $x = 0$ of system (A.1) is said to be

i) *stable if and only if* for all $t_0 \geq 0$ and for all $t \geq t_0$

$$\forall \epsilon > 0 \quad \exists \delta(\epsilon, t_0) > 0 : |x_0| < \delta(\epsilon, t_0) \Rightarrow |\varphi(t, t_0, x_0)| < \epsilon;$$

ii) *uniformly stable if and only if* in i) δ does not depend on t_0;

iii) *uniformly attractive if and only if* for all $t_0 \geq 0$

$$\exists \eta > 0 : |x_0| < \eta \Rightarrow \lim_{t \to \infty} |\varphi(t, t_0, x_0)| = 0$$

uniformly with respect to t_0 and x_0;

iv) *uniformly asymptotically stable if and only if* it is uniformly stable and uniformly attractive;

v) *unstable if and only if* it is not stable. ◇

The above definitions apply to linear time-varying systems as well as to general non-linear systems. On the other hand, when linear systems are dealt with, the equilibrium $x = 0$ is attractive *iff* it is *globally* attractive, that is definition iii) holds for all $x_0 \in \mathbb{R}^n$; this is a direct consequence of (A.4).

From this fact it follows that the property of uniform asymptotic stability, when possessed by a linear system, is always global. To this regard, note that for linear systems, as shown in Theorem A.8 below, uniform asymptotic stability only depends on the state-transition matrix.

Referring to linear systems, in this book we shall often write, with slight abuse of language, "system (A.1) is stable (uniformly stable, etc.)" in place of "the equilibrium point $x = 0$ of system (A.1) is stable (uniformly stable, etc.)."

The next theorem provides a necessary and sufficient condition for uniform asymptotic stability in terms of the state-transition matrix.

Theorem A.8 System (A.1) is uniformly asymptotically stable if and only if both the following conditions hold

i) There exists a scalar $k > 0$, such that, for all $t_0 \geq 0$, and for all $t \geq t_0$,

$$|\Phi(t, t_0)| \leq k;$$

ii)

$$\lim_{t \to \infty} |\Phi(t, t_0)| = 0$$

uniformly with respect to t_0. ▲

A.4 Input to State and Input to Output Response

Let us consider the non-autonomous LTV system (1.8); the following definition concerns the concept of controllability of the pair $(A(\cdot), F(\cdot))$.

Definition A.3 (Controllability of the pair $(A(\cdot), F(\cdot))$) Given system (1.8a), the pair $(A(\cdot), F(\cdot))$ is said to be *controllable* on $[t_0, t_1]$ if and only if, for any initial state $x_0 = x(t_0)$ and any final state $x_1 = x(t_1)$, there exists an input signal $w(\cdot)$ defined over the closed time interval $[t_0, t_1]$ that steers the state from x_0 to x_1. ◇

Related to the definition of controllability, we can introduce the Reachability Gramian associated to system (1.8).

Definition A.4 (Reachability Gramian of a LTV System) The *Reachability Gramian* of system (1.8) is defined as

$$W_r(t, t_0) := \int_{t_0}^{t} \Phi(t, \tau) F(\tau) F^T(\tau) \Phi^T(t, \tau) d\tau, \forall t \in \Omega.$$

◇

Note that, by definition the Reachability Gramian $W_r(t, t_0)$ is a symmetric and positive semidefinite matrix-valued function for all $t \geq t_0$.

Given the two Definitions A.3 and A.4, the following remark is in place.

Remark A.1 If the pair $(A(\cdot), F(\cdot))$ is *controllable*, then $W_r(t, t_0)$ is positive definite for all $t > t_0$. Furthermore, if system (1.8) is time-invariant, then $W_r(t, t_0) = W_r(t - t_0)$ and (see [93])

$$W_r(t_2 - t_0) \geq W_r(t_1 - t_0), \quad t_2 \geq t_1 \geq t_0.$$

$\diamond$

Moreover, the following lemma holds for LTV systems.

Lemma A.1 ([93]) Given system (1.8), $W_r(t, t_0)$ is the unique solution of the DLE

$$\dot{W}_r(t, t_0) = A(t)W_r(t, t_0) + W_r(t, t_0)A^T(t) + F(t)F^T(t), \tag{A.19a}$$

$$W_r(t_0, t_0) = 0. \tag{A.19b}$$

▲

Finally, denoted by $\delta(t)$ the Dirac delta function, and by $\delta_{-1}(t)$ the Heaviside step function, the impulsive response of system (1.8) is given by

$$H(t, \tau) = C(t)\Phi(t, \tau)F(\tau)\delta_{-1}(t - \tau) + G(\tau)\delta(t - \tau). \tag{A.20}$$

B

Schur Complements

In the following, we present a fundamental result of LMIs theory that is often used in this book. Such results allow to transform a Riccati-type inequality into an LMI.

First note that we have

$$\begin{pmatrix} Q & S \\ S^T & R \end{pmatrix} = \begin{pmatrix} I & 0 \\ S^T Q^{-1} & I \end{pmatrix} \begin{pmatrix} Q & 0 \\ 0 & R - S^T Q^{-1} S \end{pmatrix} \begin{pmatrix} I & Q^{-1} S \\ 0 & I \end{pmatrix}. \tag{B.1}$$

Since the left and right multipliers at the right-hand side in (B.1) are full-rank matrices, the left-hand side in (B.1) is positive definite *iff* the matrix

$$\begin{pmatrix} Q & 0 \\ 0 & R - S^T Q^{-1} S \end{pmatrix} \tag{B.2}$$

is positive definite. Therefore, we can state the following result.

Theorem B.1 The matrix

$$\begin{pmatrix} Q & S \\ S^T & R \end{pmatrix} \tag{B.3}$$

is positive definite *iff* Q is positive definite and $R - S^T Q^{-1} S$ is positive definite. ▲

The matrix $R - S^T Q^{-1} S$ is called the *Schur complement* of Q.

In the same way we can write

$$\begin{pmatrix} Q & S \\ S^T & R \end{pmatrix} = \begin{pmatrix} I & SR^{-1} \\ 0 & I \end{pmatrix} \begin{pmatrix} Q - SR^{-1} S^T & 0 \\ 0 & R \end{pmatrix} \begin{pmatrix} I & 0 \\ R^{-1} S^T & I \end{pmatrix}, \tag{B.4}$$

from which we obtain the following alternative condition for the positive definiteness of matrix (B.3).

Theorem B.2 The matrix

$$\begin{pmatrix} Q & S \\ S^T & R \end{pmatrix}$$

is positive definite *iff* R is positive definite and $Q - SR^{-1} S^T$ is positive definite. ▲

The matrix $Q - SR^{-1} S^T$ is called the Shur complement of R.

Finite-Time Stability: An Input-Output Approach, First Edition.
Francesco Amato, Gianmaria De Tommasi, and Alfredo Pironti.
© 2018 John Wiley & Sons Ltd. Published 2018 by John Wiley & Sons Ltd.

Since

$$\begin{pmatrix} Q & S \\ S^T & R \end{pmatrix} < 0 \tag{B.5}$$

can be equivalently rewritten

$$\begin{pmatrix} -Q & -S \\ -S^T & -R \end{pmatrix} > 0, \tag{B.6}$$

we have the following conditions concerning negative definiteness.

Theorem B.3 The following conditions are equivalent to each other

i) Matrix

$$\begin{pmatrix} Q & S \\ S^T & R \end{pmatrix}$$

is negative definite;

ii) both matrices Q and $R - S^T Q^{-1} S$ are negative definite;

iii) both matrices R and $Q - SR^{-1}S^T$ are negative definite. ▲

C

Computation of Feasible Solutions to Optimizations Problems Involving DLMIs

In this section we shall show how a feasibility problem constrained by a coupled DLMI/LMI or a coupled D/DLMI/LMI can be numerically solved.

C.1 Numerical Solution to a Feasibility Problem Constrained by a DLMI Coupled with LMIs

The problem we consider in this section is precisely defined as follows.

Problem C.1   Find a piecewise continuously differentiable,[1] positive definite matrix-valued function $X(\cdot)$, defined over a closed interval $\Omega := [t_0, t_0 + T]$, such that the coupled DLMI/LMI

$$\mathcal{F}(\dot{X}(t), X(t)) < 0, \quad t \in \Omega$$
$$\mathcal{G}(X(t)) < 0, \quad t \in \Omega,$$

where $\mathcal{F}(\cdot, \cdot)$ and $\mathcal{G}(\cdot)$ are assigned linear functionals of their arguments, are satisfied. ◇

In order to solve Problem C.1, we fix a sampling time T_s, and divide the interval Ω into $n = T/T_s$ sub-intervals; then the matrix-function $X(\cdot)$ is approximated by an affine function in each sub-interval.

In particular, let us consider the $(i+1)$-th sub-interval $[iT_s, (i+1)T_s]$; it is assumed that, in this time interval, the matrix-function $X(\cdot)$ takes the following form

$$X(t) = X_i + \Theta_i(t - iT_s), \qquad t \in [iT_s, (i+1)T_s), \tag{C.1}$$

where $T_s \ll T$, and X_i, Θ_i, with $i = 0, \ldots, n-1$, are the optimization variables. Note that (C.1) defines a piecewise affine matrix function $X(\cdot)$; in order to guarantee continuity of $X(\cdot)$, we must have

$$X_i + \Theta_i T_s = X_{i+1}, \quad i = 0, \ldots, n-2. \tag{C.2}$$

---

1 Remember that a piecewise continuously differentiable function in a closed interval Ω is: i) continuous over Ω, and ii) piecewise differentiable in Ω, i.e., the interval Ω can be divided into a finite number of sub-intervals, and in each sub-interval the function is differentiable.

Finite-Time Stability: An Input-Output Approach, First Edition.
Francesco Amato, Gianmaria De Tommasi, and Alfredo Pironti.
© 2018 John Wiley & Sons Ltd. Published 2018 by John Wiley & Sons Ltd.

It is straightforward to recognize that such a piecewise affine matrix-function can approximate a general continuous matrix-function $P(\cdot)$ with adequate accuracy, provided that T_s is sufficiently small.

From (C.1) it follows that, for $t \in [iT_s, (i+1)T_s)$,

$$\dot{X}(t) = \Theta_i, \quad i = 0, \dots, n-1. \tag{C.3}$$

Replacing the expression of $X(\cdot)$ and its derivative, namely (C.1) and (C.3), into the constraints of Problem C.1, we have, for $t \in [iT_s, (i+1)T_s)$,

$$\mathcal{F}(\dot{X}(t), X(t)) = \mathcal{F}(\Theta_i, X_i + \Theta_i(t - iT_s)) \tag{C.4a}$$

$$\mathcal{G}(X(t)) = \mathcal{G}(X_i + \Theta_i(t - iT_s)). \tag{C.4b}$$

Since $\mathcal{F}(\cdot, \cdot)$ and $\mathcal{G}(\cdot)$ are affine functions of their arguments, and $\Theta_i(t - iT_s)$ is an affine function of t, we have that the right-hand side of both (C.4a) and (C.4b) are affine functions of t. Therefore, according to [96, Appendix A1], $\mathcal{F}(\dot{X}(\cdot), X(\cdot))$ and $\mathcal{G}(X(\cdot))$ are negative definite if and only if they are negative definite at the extrema of the interval. Hence, taking into account (C.2),

$$\mathcal{F}(\dot{X}(t), X(t)) < 0, \quad t \in [iT_s, (i+1)T_s) \;\Leftrightarrow\; \mathcal{F}(\Theta_i, X_i) < 0, \quad \mathcal{F}(\Theta_i, X_{i+1}) < 0 \tag{C.5a}$$

$$\mathcal{G}(X(t)) < 0, \quad t \in [iT_s, (i+1)T_s) \;\Leftrightarrow\; \mathcal{G}(X_i) < 0, \quad \mathcal{G}(X_{i+1}) < 0. \tag{C.5b}$$

Finally, again we have that the affine matrix-valued function (C.1) is positive definite in the interval $[iT_s, (i+1)T_s]$ if and only if it is positive definite at the extrema of the interval, that is

$$X_i + \Theta_i(t - iT_s) > 0, \qquad t \in [iT_s, (i+1)T_s), \tag{C.6}$$

if and only if

$$X_i > 0 \tag{C.7a}$$

$$X_i + \Theta_i T_s = X_{i+1} > 0. \tag{C.7b}$$

Taking into account (C.2), (C.5) and (C.7), we have that, at the price of some conservatism, depending on how small the sampling time T_s is, the feasibility Problem C.1 can be replaced by the following problem, subject to classical LMIs constraints.

Problem C.2 Find symmetric matrices X_i, Θ_i, with $i = 0, \dots, n-1$, such that

$$X_i > 0$$

$$\mathcal{F}(\Theta_i, X_i) < 0, \quad \mathcal{F}(\Theta_i, X_{i+1}) < 0$$

$$\mathcal{G}(X_i) < 0, \quad \mathcal{G}(X_{i+1}) < 0$$

$$X_i + \Theta_i T_s = X_{i+1},$$

for $i = 0, \dots, n-1$. ◇

Problem C.2 can be solved by exploiting standard optimization tools such as the MAT-LAB LMI Toolbox® ([90]).

C.2 Numerical Solution to a Feasibility Problem Constrained by a D/DLMI Coupled with LMIs

The problem we consider in this section is precisely defined as follows: consider the usual finite-time interval $\Omega := [t_0, t_0 + T]$, and the finite set

$$\mathcal{T} := \{t_1, \dots, t_v\} \subset \Omega, \quad t_1 > t_0, \quad t_v < t_0 + T.$$

Problem C.3 Find a piecewise continuous positive definite matrix-valued function $X(\cdot)$, defined over the closed interval Ω, such that the coupled D/DLMI/LMI

$$\mathcal{F}(\dot{X}(t), X(t)) < 0, \quad t \in \Omega, \quad t \notin \mathcal{T}$$
$$\mathcal{D}(X^+(t_k), X(t_k)) < 0, \quad t_k \in \mathcal{T}$$
$$\mathcal{G}(X(t)) < 0, \quad t \in \Omega,$$

where $\mathcal{F}(\cdot, \cdot)$, $\mathcal{D}(\cdot, \cdot)$ and $\mathcal{G}(\cdot)$ are assigned linear functionals of their arguments, are satisfied. ◇

In order to solve Problem C.3, let us consider the interval between two resetting times t_k and t_{k+1}. In this time interval, it is assumed that the matrix-function $X(\cdot)$ has the following piecewise affine structure

$$X(t) = \begin{cases} X_k + \Theta_{k,1}(t - t_k), & t \in [t_k, t_k + T_s) \\ X_k + \sum_{i=1}^{j} \Theta_{k,i} T_s + \Theta_{k,j+1}(t - jT_s - t_k) \\ \qquad t \in [t_k + jT_s, t_k + (j+1)T_s), \\ \qquad j = 1, \dots, \mathcal{I}_k - 1, \\ X_k + \sum_{i=1}^{\mathcal{I}_k} \Theta_{k,i} T_s + \Theta_{k, \mathcal{I}_k+1}(t - \mathcal{I}_k T_s - t_k) \\ \qquad t \in [t_k + \mathcal{I}_k T_s, t_{k+1}), \end{cases} \tag{C.8}$$

where $\mathcal{I}_k := \max\{j \in \mathbb{N} : j < (t_{k+1} - t_k)/T_s\}$, $T_s \ll T$, and X_k, $\Theta_{k,i}$, are the optimization variables.

Note that (C.8) guarantees continuity of $X(\cdot)$ inside the interval $[t_k, t_{k+1})$.

Furthermore, in correspondence of a given resetting time t_k, the matrix-function $X(\cdot)$ jumps between

$$X_{k-1} + \sum_{i=1}^{\mathcal{I}_{k-1}} \Theta_{k-1,i} T_s + \Theta_{k-1, \mathcal{I}_{k-1}+1}(t_k - \mathcal{I}_{k-1} T_s - t_{k-1})$$

and X_k.

From (C.8), it follows that, in the time interval $[t_k, t_{k+1})$,

$$\dot{X}(t) = \begin{cases} \Theta_{k,j+1}, & t \in [t_k + jT_s, t_k + (j+1)T_s), \quad j = 0, \dots, \mathcal{I}_k - 1 \\ \Theta_{k, \mathcal{I}_k+1}, & t \in [t_k + \mathcal{I}_k T_s, t_{k+1}). \end{cases} \tag{C.9}$$

Replacing the expression of $X(\cdot)$ and its derivative, namely (C.8) and (C.9), into the constraints of Problem C.3, we have, for $t \in [t_k, t_{k+1})$,

$$
\mathcal{F}(\dot{X}(t), X(t)) = \begin{cases}
F(\Theta_{k,1}, X_k + \Theta_{k,1}(t - t_k)), & t \in [t_k, t_k + T_s) \\
F\left(\Theta_{k,j+1}, X_k + \sum_{i=1}^{j} \Theta_{k,i} T_s + \Theta_{k,j+1}(t - jT_s - t_k)\right) \\
\quad t \in [t_k + jT_s, t_k + (j+1)T_s), \quad j = 1, \ldots, I_k - 1 \\
F\left(\Theta_{k,I_k+1}, X_k + \sum_{i=1}^{I_k} \Theta_{k,i} T_s + \Theta_{k,I_k+1}(t - I_k T_s - t_k)\right) \\
\quad t \in [t_k + I_k T_s, t_{k+1}),
\end{cases}
$$

(C.10a)

$$
D(X^+(t_k), X(t_k)) = D\left(X_k, X_{k-1} + \sum_{i=1}^{I_{k-1}} \Theta_{k-1,i} T_s + \Theta_{k-1,I_{k-1}+1}(t_k - I_{k-1} T_s - t_{k-1})\right)
$$

(C.10b)

$$
\mathcal{G}(X(t)) = \begin{cases}
G(X_k + \Theta_{k,1}(t - t_k)), & t \in [t_k, t_k + T_s) \\
G\left(X_k + \sum_{i=1}^{j} \Theta_{k,i} T_s + \Theta_{k,j+1}(t - jT_s - t_k)\right) \\
\quad t \in [t_k + jT_s, t_k + (j+1)T_s), \quad j = 1, \ldots, I_k - 1 \\
G\left(X_k + \sum_{i=1}^{I_k} \Theta_{k,i} T_s + \Theta_{k,I_k+1}(t - I_k T_s - t_k)\right) \\
\quad t \in [t_k + I_k T_s, t_{k+1}).
\end{cases}
$$

(C.10c)

Following the same derivation of Section C.1, we can conclude that $\mathcal{F}(\dot{X}(\cdot), X(\cdot))$, and $\mathcal{G}(X(\cdot))$ are affine functions of t. Therefore, according to [96, Appendix A1], we have that $\mathcal{F}(\dot{X}(\cdot), X(\cdot))$ and $\mathcal{G}(X(\cdot))$ are negative definite if and only if they are negative definite at the extrema of the interval. Hence $\mathcal{F}(\dot{X}(t), X(t)) < 0$, for $t \in [t_k, t_{k+1})$, if and only if

$$
\begin{cases}
F(\Theta_{k,1}, X_k) < 0, \quad F(\Theta_{k,1}, X_k + \Theta_{k,1} T_s) < 0, \\
F\left(\Theta_{k,j+1}, X_k + \sum_{i=1}^{j} \Theta_{k,i} T_s\right) < 0, \quad F\left(\Theta_{k,j+1}, X_k + \sum_{i=1}^{j} \Theta_{k,i} T_s + \Theta_{k,j+1} T_s\right) < 0, \\
\quad\quad\quad\quad\quad\quad\quad\quad\quad\quad\quad\quad\quad\quad\quad\quad\quad\quad\quad\quad\quad\quad j = 1, \ldots, I_k - 1, \\
F\left(\Theta_{k,I_k+1}, X_k + \sum_{i=1}^{I_k} \Theta_{k,i} T_s\right) < 0, \\
F\left(\Theta_{k,I_k+1}, X_k + \sum_{i=1}^{I_k} \Theta_{k,i} T_s + \Theta_{k,I_k+1}(t_{k+1} - I_k T_s - t_k)\right) < 0,
\end{cases}
$$

(C.11)

and, in the same way, $\mathcal{G}(X(t)) < 0$, for $t \in [t_k, t_{k+1})$, if and only if

$$
\begin{cases}
G(X_k) < 0, \quad G(X_k + \Theta_{k,1} T_s) < 0 \\
G\left(X_k + \sum_{i=1}^{j} \Theta_{k,i} T_s\right) < 0, \quad G\left(X_k + \sum_{i=1}^{j} \Theta_{k,i} T_s + \Theta_{k,j+1} T_s\right) < 0 \\
\quad\quad\quad\quad\quad\quad\quad\quad\quad\quad\quad\quad\quad\quad\quad\quad j = 1, \ldots, I_k - 1, \\
G\left(X_k + \sum_{i=1}^{I_k} \Theta_{k,i} T_s\right) < 0, \\
G\left(X_k + \sum_{i=1}^{I_k} \Theta_{k,i} T_s + \Theta_{k,I_k+1}(t_{k+1} - I_k T_s - t_k)\right) < 0.
\end{cases}
$$

(C.12)

Moreover, note that the matrix-valued function (C.8) is piecewise affine and contin-uous in the interval $[t_k, t_{k+1})$; therefore, it is positive definite in the same interval if and only if

$$\begin{cases} X_k > 0, \quad X_k + \Theta_{k,1}T_s > 0 \\ X_k + \sum_{i=1}^{j} \Theta_{k,i}T_s > 0, \quad X_k + \sum_{i=1}^{j} \Theta_{k,i}T_s + \Theta_{k,j+1}T_s > 0 \\ \qquad\qquad\qquad\qquad\qquad j = 1, \dots, I_k - 1, \qquad\qquad\text{(C.13)} \\ X_k + \sum_{i=1}^{I_k} \Theta_{k,i}T_s > 0, \\ X_k + \sum_{i=1}^{I_k} \Theta_{k,i}T_s + \Theta_{k,I_k+1}(t_{k+1} - I_k T_s - t_k) > 0. \end{cases}$$

Taking into account (C.11), (C.10b), (C.12), and (C.13), we have that, at the price of some conservatism, the feasibility Problem C.3 can be replaced by the following prob-lem, subject to classical LMIs constraints.

Problem C.4   Find symmetric matrices $X_k, \Theta_{k,i}$, with $i = 1, \dots, I_k + 1$ and $k = 1, \dots, v$, such that

$$\begin{cases} \mathcal{F}(\Theta_{k,1}, X_k) < 0, \quad \mathcal{F}(\Theta_{k,1}, X_k + \Theta_{k,1}T_s) < 0 \\ \mathcal{F}\left(\Theta_{k,j+1}, X_k + \sum_{i=1}^{j} \Theta_{k,i}T_s\right) < 0, \quad \mathcal{F}\left(\Theta_{k,j+1}, X_k + \sum_{i=1}^{j} \Theta_{k,i}T_s + \Theta_{k,j+1}T_s\right) < 0 \\ \qquad\qquad\qquad\qquad\qquad j = 1, \dots, I_k - 1 \\ \mathcal{F}\left(\Theta_{k,I_k+1}, X_k + \sum_{i=1}^{I_k} \Theta_{k,i}T_s\right) < 0, \\ \mathcal{F}\left(\Theta_{k,I_k+1}, X_k + \sum_{i=1}^{I_k} \Theta_{k,i}T_s + \Theta_{k,I_k+1}(t_{k+1} - I_k T_s - t_k)\right) < 0, \\ \mathcal{D}\left(X_k, X_{k-1} + \sum_{i=1}^{I_{k-1}} \Theta_{k-1,i}T_s + \Theta_{k-1,I_{k-1}+1}(t_k - I_{k-1}T_s - t_{k-1})\right) < 0 \\ \mathcal{G}(X_k) < 0, \quad \mathcal{G}(X_k + \Theta_{k,1}T_s) < 0 \\ \mathcal{G}\left(X_k + \sum_{i=1}^{j} \Theta_{k,i}T_s\right) < 0, \quad \mathcal{G}\left(X_k + \sum_{i=1}^{j} \Theta_{k,i}T_s + \Theta{k,j+1}T_s\right) < 0 \\ \qquad\qquad\qquad\qquad\qquad j = 1, \dots, I_k - 1, \\ \mathcal{G}\left(X_k + \sum_{i=1}^{I_k} \Theta_{k,i}T_s\right) < 0, \\ \mathcal{G}\left(X_k + \sum_{i=1}^{I_k} \Theta_{k,i}T_s + \Theta_{k,I_k+1}(t_{k+1} - I_k T_s - t_k)\right) < 0, \\ X_k > 0, \quad X_k + \Theta_{k,1}T_s > 0 \\ X_k + \sum_{i=1}^{j} \Theta_{k,i}T_s > 0, \quad X_k + \sum_{i=1}^{j} \Theta_{k,i}T_s + \Theta_{k,j+1}T_s > 0 \\ \qquad\qquad\qquad\qquad\qquad j = 1, \dots, I_k - 1, \\ X_k + \sum_{i=1}^{I_k} \Theta_{k,i}T_s > 0, \\ X_k + \sum_{i=1}^{I_k} \Theta_{k,i}T_s + \Theta_{k,I_k+1}(t_{k+1} - I_k T_s - t_k) > 0. \end{cases}$$

$$\text{(C.14)}$$

for $k = 1, \dots, v$. $\qquad\qquad\qquad\qquad\qquad\qquad\qquad\qquad\qquad\qquad\qquad\qquad\quad\diamond$

Again, Problem C.4 can be solved exploiting standard optimization tools such as the MATLAB LMI Toolbox® ([90]).

D

Solving Optimization Problems Involving DLMIs using MATLAB®

This appendix presents the code used to solve the feasibility problems for some of the examples of this book. In particular, two MATLAB® scripts are presented; the former solves an optimization problem with DLMI/LMI constraints, while the latter deals with a D/DLMI/LMI feasibility problem. Both scripts make use of the YALMIP parser [89], which allows the programmer to specify constrained optimization problems in a very intuitive way.

D.1 MATLAB® Script for the Solution of the Optimization Problem with DLMI/LMI Constraints Presented in Example 2.2

The MATLAB® script reported in Listing D.1 solves the optimization problem with the DLMI/LMI constraints described in Example 2.2. This optimization problem is based on the sufficient condition to verify IO-FTS with respect to $(\Omega, \mathcal{W}_\infty, Q(\cdot))$ given in Corollary 2.3.

The proposed implementation is based on the approximation presented in Section C.1 of Appendix C. In order to implicitly satisfy the equality constraints (C.2), taking into account that in Example 2.2 the considered finite-time interval is equal to $\Omega = [0, T]$, with $T = 1.5$ s, the piecewise continuously differentiable matrix-valued function $P(\cdot)$ is specified as a piecewise linear function, i.e.,

$$P(t) = \begin{cases} P_0 + \Theta_1 t, & t \in [0, T_s), \\ P_0 + \sum_{h=1}^{j} \Theta_h T_s + \Theta_{j+1}(t - jT_s), \\ \qquad t \in [jT_s, (j+1)T_s) \\ \qquad j = 1, \dots, n, \end{cases} \tag{D.1}$$

where $T_s = 0.02$ s is the sampling time used for the discretization of the DLMI/LMI constraints (2.45) specified in Corollary 2.3, and $n = T/T_s$ is the number of subintervals. It is straightforward to notice that, if $P(\cdot)$ is specified as in D.1, then continuity among two contiguous subintervals is guaranteed.

In the code reported in Listing D.1, the optimization variables are defined at lines 22–25 using the YALMIP command sdpvar. In particular, the symmetric matrices Θ_i are dynamically defined using the MATLAB command eval. Similarly, the DLMI/LMI constraints (2.45) are dynamically specified at lines 39–49. It should be

Finite-Time Stability: An Input-Output Approach, First Edition.
Francesco Amato, Gianmaria De Tommasi, and Alfredo Pironti.
© 2018 John Wiley & Sons Ltd. Published 2018 by John Wiley & Sons Ltd.

noticed that the DLMI constraints are specified for both extrema of each subinterval (see lines 43–46). Moreover, the positive definiteness of $P(\cdot)$ is implied by constraints (2.45b), and hence it does not need to be explicitly included.

The choice of the specific solver is made at line 54, where the MATLAB LMI Toolbox® is selected. Eventually, having specified -Q as objective function, when the command optimize is launched, an optimization problem that minimizes the objective function (i.e., maximizes Q) subject to the given constraints is solved, that is, the optimization problem described in Example 2.2 is solved. In particular, the maximum value for Q returned by the optimization is about 0.727.

D.2   MATLAB® Script for the Solution of the D/DLMI/LMI Feasibility Problem Presented in Section 8.3.1

In Section 8.3.1 we exploited Theorem 8.4 in order to design a state-feedback controller that IO finite-time stabilizes the IDLS described in Section 7.4.1 with respect to $(\Omega, \mathcal{W}_\infty, Q)$.

In the code reported in Listing D.2, the piecewise affine structure (C.8) is adopted to approximate the matrix-valued function $\Upsilon(\cdot)$ specified in Theorem 8.4, while the matrix-valued function $L(\cdot)$ is assumed piecewise linear.

The parameters of the considered IDLS are specified in Listing D.2 at lines 1–14, while at lines 38–46 the samples of the time-varying dynamic matrix are calculated.

The optimization variables used to model $\Upsilon(\cdot)$ and $L(\cdot)$ are defined at lines 48–59. Before defining the D/DLMI/LMI constraints specified in Theorem 8.4, the positive definiteness of $\Upsilon(\cdot)$ is explicitly imposed by the constraints at lines 63–73. In order to implement the constraint (8.12) the square root of the variable time is used at lines 81, 95 and 105. The feasibility problem is then solved by specifying [] as objective function, in the optimize command at line 125.

Listing D.1 MATLAB® script used to solve the optimization problem with DLMI/LMI constraints described in Example 2.2.

```
1   % Second order BIBO stable system
2   A = [0 1;-2 -3];
3   F = [1 0;0 1];
4   C = [1 0];
5   G = [0 0];
6   sys = ss(A,F,C,G);
7
8   % Input weighting matrix
9   R = eye(2);
10
11  % Finite-time interval
12  T = 1.5;
13
14  % Sampling time for the discretization of DLMI/LMI constraints
15  Ts = 2e-2;
16
```

```
17   % Number of subintervals
18   nOfIntervals = T/Ts;
19
20   % Positive definite matrices used to approximate a generic P(t)
21   % in each time interval using a piecewise linear approximation
22   P_0 = sdpvar(size(A,1));
23   for i = 1:nOfIntervals
24       eval(['Theta_' num2str(i) ' = sdpvar(' num2str(size(A,1)) ');
25           ']);
26   end
27
28   % Output weighting matrix
29   % In the considered example -Q represents the objective function
30   Q = sdpvar(1);
31
32   % Contraints W_inf (Corollary 2.3 - Chapter 2)
33   constr = [];
34
35   % P_0 should be positive definite
36   constr = [P_0>0];
37
38   time = 0;
39   PP = ['P_0'];
40   for i = 1:nOfIntervals
41       % Constraint (2.51a)
42       % The DLMI constraint is specified at both extrema
43       % of each subinterval
44       eval(['constr = constr + [[(Theta_' num2str(i) '+(A)''*(' PP
45           ')+(' PP ')*(A)) (' PP ')*F; F''*(' PP ') -R ]<0];']);
46       PP = [PP '+Theta_' num2str(i) '*Ts'];
47       time = time + Ts;
48       eval(['constr = constr + [[(Theta_' num2str(i) '+(A)''*(' PP
49           ')+(' PP ')*(A)) (' PP ')*F; F''*(' PP ') -R ]<0];']);
50       % Constraint (2.51b)
51       eval(['constr = constr + [' PP ' > C''*time*Q*C];']);
52   end
53
54   % Parameters for the LMI optimization tool
55   ops = sdpsettings;
56   % The MATLAB LMI Toolbox is selected
57   ops.solver = 'lmilab';
58   ops.showprogress = 1;
59   ops.removeequalities = 2;
60
61   % Solve feasibility problem
62   t_start = cputime;
63   out = optimize(constr,-Q,ops);
64   time_to_solve = cputime-t_start;
```

Listing D.2 MATLAB® script used to solve the D/DLMI/LMI feasibility problem used in Section 8.3.1 to design a state feedback controller.

```matlab
1   % Continuous-time dynamic
2   sys = tf(25,[1 2.5 25]);
3   [A,F,C,G] = ssdata(sys);
4   B = [1;1];
5   % Resetting law
6   J = -0.8*eye(size(A,1));
7
8   % Variance rate of the dynamic matrix
9   Rho=zeros(size(A,1));
10  Rho(1,1)=.5;
11  Rho(end,end)=.5;
12
13  % Resetting times
14  TT = [.25 .5 .75 1 1.25 1.5 1.75];
15
16  % IO-FTS parameters
17  R = 1;
18  Q = 1;
19  T = 2;
20
21  % Sampling time for discretization of D/DLMI/LMI
22  Ts = .05;
23
24  % Number of intervals
25  nOfIntervals = length(TT)+1;
26  % Number of time samples in each interval
27  nOfSamples = zeros(1,nOfIntervals);
28  tStart = 0;
29  for i = 1:nOfIntervals
30      if i == nOfIntervals
31          nOfSamples(i) = (T-tStart)/Ts;
32      else
33          nOfSamples(i) = (TT(i)-tStart)/Ts;
34          tStart = TT(i);
35      end
36  end
37
38  % Time varying dynamic matrix
39  A_0=A;
40  for i=1:sum(nOfSamples)
41      eval(['A_' num2str(i) '= A + Rho*i*Ts;']);
42  end
43  A_tv=[];
44  for i=0:sum(nOfSamples)
45          eval(['A_tv =[A_tv; A_' num2str(i) '];']);
46  end
47
48  % Optimization variables
```

```
49  % Piecewise affine approximation for \Upsilon(\cdot)
50  for i = 1:nOfIntervals
51      eval(['P_' num2str(i) ' = sdpvar(' num2str(size(A,1)) ');']);
52      for j = 1:nOfSamples(i)
53          eval(['Slope_' num2str(i) '_' num2str(j) ' = sdpvar('
54              num2str(size(A,1)) ');']);
55      end
56  end
57  % Piecewise linear approximation for L(\cdot)
58  for i = 1:(sum(nOfSamples)+1)
59      eval(['L_' num2str(i) '= sdpvar(' num2str(size(B,2)) ','
60          num2str(size(A,1)) ',''full'');']);
61  end
62
63  % Constraints
64  % Positive definitiveness of \Upsilon
65  constr = [];
66  for i = 1:nOfIntervals
67      eval(['constr = constr + [P_' num2str(i) '>0*eye(size(A,1))];
68          ']);
69      for j = 1:nOfSamples(i)
70          P = ['P_' num2str(i)];
71          for k = 1:j
72              P = [P '+Slope_' num2str(i) '_' num2str(k) '*Ts'];
73          end
74          eval(['constr = constr + [' P '>0*eye(size(' P '))];']);
75      end
76  end
77  % Constraints (8.9a) and (8.12)
78  time = 0;
79  LIndex = 1;
80  for i = 1:nOfIntervals
81      % First time instant - constraints (8.9a)
82      eval(['constr = constr + [[-Slope_' num2str(i) '_1+(P_'
83          num2str(i) ')*A_' num2str(i) '''+A_' num2str(i) '*(P_'
84          num2str(i) ')+B*L_' num2str(LIndex) '+L_' num2str(LIndex)
85          '''*B'' F;'...
86          'F'' -R]<0];']);
87      % Constraints (8.12)
88      eval(['constr = constr + [[P_' num2str(i) ' sqrt(' num2str(
89          time) ')*P_' num2str(i) '*C'';'...
90          'sqrt(' num2str(time) ')*C*P_' num2str(i) ' inv(Q)]>0];'
91          ]);
92      time = time + Ts;
93      LIndex = LIndex + 1 ;
94      P = ['P_' num2str(i)];
95      for j = 1:nOfSamples(i)-1
96          P = [P '+Slope_' num2str(i) '_' num2str(j) '*Ts'];
97          % Left derivative - constraints (8.9a)
98          eval(['constr = constr + [[-Slope_' num2str(i) '_'
99              num2str(j) '+(' P ')*A_' num2str(i+j) '''+A_' num2str
```

```
100                    (i+j) '*(' P ')+B*L_' num2str(LIndex) '+L_' num2str(
101                    LIndex) '''*B'' F;'...
102                    'F'' -R]<0];']);
103           % Right derivative - constraints (8.9a)
104           eval(['constr = constr + [[-Slope_' num2str(i) '_'
105                    num2str(j+1) '+(' P ')*A_' num2str(i+j) '''+A_'
106                    num2str(i+j) '*(' P ')+B*L_' num2str(LIndex) '+L_'
107                    num2str(LIndex) '''*B'' F;'...
108                    'F'' -R]<0];']);
109           % Constraints (8.12)
110           eval(['constr = constr + [[' P ' sqrt(' num2str(time) ')
111                    *(' P ')*C'';'...
112                    'sqrt(' num2str(time) ')*C*(' P ') inv(Q)]>0];']);
113           time = time + Ts;
114           LIndex = LIndex + 1;
115       end
116       % Last time instant - constraints (8.9a)
117       P = [P '+Slope_' num2str(i) '_' num2str(nOfSamples(i)) '*Ts'
118           ];
119       eval(['constr = constr + [[-Slope_' num2str(i) '_' num2str(
120           nOfSamples(i)) '+(' P ')*A_' num2str(i+nOfSamples(i)) '''
121           +A_' num2str(i+nOfSamples(i)) '*(' P ')+B*L_' num2str(
122           LIndex) '+L_' num2str(LIndex) '''*B'' F;'...
123           'F'' -R]<0];']);
124       % Constraints (8.12)
125       eval(['constr = constr + [[' P ' sqrt(' num2str(time) ')*(' P
126           ')*C'';'...
127           'sqrt(' num2str(time) ')*C*(' P ') inv(Q)]>0];']);
128   end
129   % Contraints (8.9b)
130   for i = 1:length(TT)
131       P = ['P_' num2str(i)];
132       for j = 1:nOfSamples(i)
133           P = [P '+Slope_' num2str(i) '_' num2str(j) '*Ts'];
134       end
135       eval(['constr = constr + [[' P ' (' P ')*J'';J*(' P ') P_'
136           num2str(i+1) ']>0];']);
137   end
138
139   % Solve feasibility problem
140   ops = sdpsettings;
141   ops.solver = 'lmilab';
142   ops.showprogress = 1;
143   ops.removeequalities = 2;
144   optimize(constr,[],ops);
```

E

Examples of Applications of IO-FTS Control Design to Real-World Systems

E.1 Building Subject to Earthquakes

Let us consider an N-story building subject to an earthquake. The building lumped parameters model is reported in Figure E.1. The control system is made by a base isolator together with an actuator that generates a control force on the base floor.

The aim of the isolator is to produce a dynamic decoupling of the structure from its foundation. If this is the case, the inter-story drifts are reduced, and the building behavior can be approximated by the one of a rigid body [131]. Furthermore, the description of the system in terms of absolute coordinates, i.e., when the displacement is defined with respect to an inertial reference frame, ensures that the disturbances act only at the base floor [132].

It turns out that it is sufficient to provide an actuator only on the base floor in order to keep the displacement and velocity of the structure under a specified boundary. Indeed, the goal of the control system is to overcome the forces generated by the isolation system at the base floor, in order to minimize the absolute displacement and velocity of the structure.

The state-space model of the considered system is

$$\dot{x}(t) = Ax(t) + Bu(t) + Gw(t) \tag{E.1a}$$
$$y(t) = Cx(t) \tag{E.1b}$$

If we denote by $s_0(\cdot)$ and $\dot{s}_0(\cdot)$ the displacement and the velocity of the ground and with $s_i(\cdot)$ and $\dot{s}_i(\cdot)$ the displacement and the velocity of the i-th floor, then the state vector can be defined as $x(\cdot) = [x_1(\cdot) \; x_2(\cdot) \; \dots \; x_{2N}(\cdot)]^T$, where $x_i(\cdot) = \dot{s}_i(\cdot)$ and $x_{i+N}(\cdot) = s_i(\cdot)$, $i = 1, \dots, N$. The vector $w(\cdot) = [s_0(\cdot) \; \dot{s}_0^T(\cdot)]$ represents the exogenous input and $u(t)$ is the control force applied to the base floor. The model matrices in (E.1) are

$$A = \begin{pmatrix} A_1 & A_2 \\ I & 0 \end{pmatrix}, \quad B = \begin{pmatrix} 1/m_1 \\ 0 \end{pmatrix}, \quad G = \begin{pmatrix} k_0/m_1 & c_0/m_1 \\ 0 & 0 \end{pmatrix},$$

$$C = \begin{pmatrix} \dfrac{-(c_0 + c_1)}{m_1} & \dfrac{c_1}{m_1} & \underbrace{0 \; \cdots \; 0}_{N-2} & \dfrac{-(k_0 + k_1)}{m_1} & \dfrac{k_1}{m_1} & \underbrace{0 \; \cdots \; 0}_{N-2} \end{pmatrix},$$

Finite-Time Stability: An Input-Output Approach, First Edition.
Francesco Amato, Gianmaria De Tommasi, and Alfredo Pironti.
© 2018 John Wiley & Sons Ltd. Published 2018 by John Wiley & Sons Ltd.

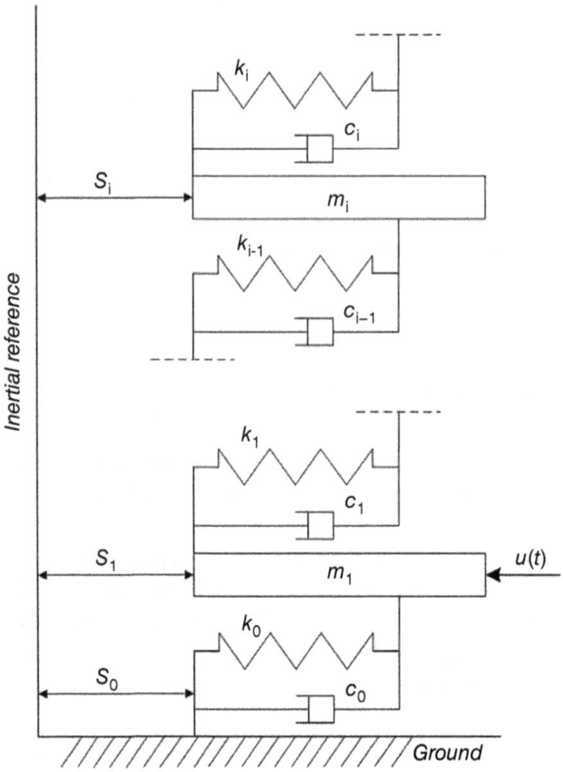

Figure E.1 Lumped parameters model of an N-story building.

where A_1 and A_2 are $N \times N$ tridiagonal matrices defined as

$$
A_1 = \begin{pmatrix}
-\dfrac{(c_0 + c_1)}{m_1} & \dfrac{c_1}{m_1} & 0 & & \cdots & & 0 \\
& & \cdots\;\cdots & & & & \\
0 & \cdots & \dfrac{c_{i-1}}{m_i} & -\dfrac{(c_{i-1} + c_i)}{m_i} & \dfrac{c_i}{m_i} & \cdots & 0 \\
& & & \cdots\;\cdots & & & \\
0 & & \cdots & & 0 & \dfrac{c_{N-1}}{m_N} & -\dfrac{c_{N-1}}{m_N}
\end{pmatrix},
$$

$$
A_2 = \begin{pmatrix}
-\dfrac{(k_0 + k_1)}{m_1} & \dfrac{k_1}{m_1} & 0 & & \cdots & & 0 \\
& & \cdots\;\cdots & & & & \\
0 & \cdots & \dfrac{k_{i-1}}{m_i} & -\dfrac{(k_{i-1} + k_i)}{m_i} & \dfrac{k_i}{m_i} & \cdots & 0 \\
& & & \cdots\;\cdots & & & \\
0 & & \cdots & & 0 & \dfrac{k_{N-1}}{m_N} & -\dfrac{k_{N-1}}{m_N}
\end{pmatrix}.
$$

The model parameters are reported in Table E.1 for the six-story building considered in this example.

Taking into account the presence of the isolator and given the choice of the C matrix, the controlled output is related to the acceleration at the ground floor. Concerning the

Table E.1 Model parameters for the considered six story building (N=6).

Floor #	Mass [kg]	Spring coefficient [kN/m]	Damping coefficient [kNs/m]
0	–	$k_0=1200$	$c_0=2.4$
1	$m_1=6800$	$k_1=33732$	$c_1=67$
2	$m_2=5897$	$k_2=29093$	$c_2=58$
3	$m_3=5897$	$k_3=28621$	$c_3=57$
4	$m_4=5897$	$k_4=24954$	$c_4=50$
5	$m_5=5897$	$k_5=19059$	$c_5=38$
6	$m_6=5897$	–	–

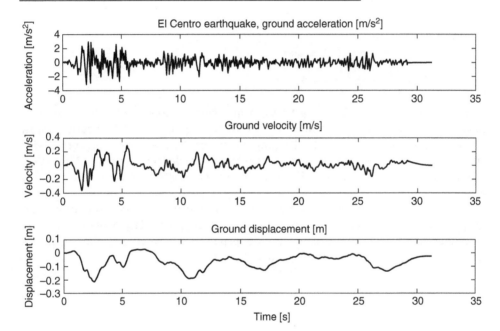

Figure E.2 Ground acceleration, velocity, and displacement of El Centro earthquake.

choice of the IO-FTS parameters, for a given geographic area these can be chosen starting from the worst earthquake on record. Indeed, from the time trace of the ground acceleration, velocity, and displacement of the *El Centro* earthquake (May 18, 1940) reported in Figure E.2, the following IO-FTS parameters have been considered

$$R = I, \quad Q = 0.1, \quad \Omega = [0, 35]. \tag{E.2}$$

E.2 Quarter Car Suspension Model

Here, we consider a typical engineering case study, namely, a vehicle active suspension system; as we shall see, the typical constraints that arise in this application field can be effectively framed in the IO-FTS context.

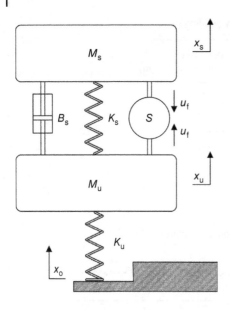

Figure E.3 Schematic representation of the active suspension system.

The scheme of a two-degree-of-freedom quarter-car model, taken from [102], is reported in Figure E.3: the system comprises the sprung mass, M_s, the unsprung mass, M_u, the suspension damper with damping coefficient B_s, the suspension spring with elastic coefficient K_s, the elastic effect caused by the tire deflection, modeled by means of a spring with elastic coefficient K_u, the hydraulic actuator S, generating a scalar active force u_f.

In the following we denote by x_s and x_u the vertical displacements of the sprung and unsprung masses, respectively, and by x_o the vertical ground displacement caused by the road unevenness. The state variables are the suspension stroke $x_s - x_u$, the tire deflection $x_u - x_o$, and the derivatives of x_s and x_u, that is,

$$x_1 = x_s - x_u$$
$$x_2 = \dot{x}_s$$
$$x_3 = x_u - x_o$$
$$x_4 = \dot{x}_u.$$

The resulting open-loop dynamical model reads

$$\dot{x}(t) = \begin{pmatrix} 0 & 1 & 0 & -1 \\ -\dfrac{K_s}{M_s} & -\dfrac{B_s}{M_s} & 0 & \dfrac{B_s}{M_s} \\ 0 & 0 & 0 & 1 \\ \dfrac{K_s}{M_u} & \dfrac{B_s}{M_u} & -\dfrac{K_u}{M_u} & -\dfrac{B_s}{M_u} \end{pmatrix} x(t) + \begin{pmatrix} 0 \\ \dfrac{u_{\max}}{M_s} \\ 0 \\ -\dfrac{u_{\max}}{M_u} \end{pmatrix} u(t) + \begin{pmatrix} 0 \\ 0 \\ -1 \\ 0 \end{pmatrix} w(t), \qquad \text{(E.3)}$$

where the normalized active force $u(\cdot) = u_f(\cdot)/u_{\max}$ is the control input, and $w(\cdot) = \dot{x}_o(\cdot)$ represents the disturbance caused by the road roughness.

When designing a controller for an active suspension system, a number of constraints should be considered ([102]). In order to ensure a firm uninterrupted contact of the wheels to the road, the dynamic tire load should not exceed the static one, that is,

$$K_u|x_3(t)| < (M_s + M_u)g, \qquad t \geq 0, \tag{E.4}$$

and the suspension stroke should fulfill the following constraint

$$|x_1(t)| \leq SS, \qquad t \geq 0, \tag{E.5}$$

where SS is the maximum allowable value for the suspension stroke. Therefore, in order to cast the control design problem into the IO-FTS framework, we consider the following system outputs

$$\begin{pmatrix} y_1(t) \\ y_2(t) \\ y_3(t) \end{pmatrix} = \begin{pmatrix} \dot{x}_2(t) \\ \dfrac{x_1(t)}{SS} \\ \dfrac{K_u x_3(t)}{g(M_s + M_u)} \end{pmatrix} = \begin{pmatrix} C_1 \\ C_2 \\ C_3 \end{pmatrix} x(t) + \begin{pmatrix} D_1 \\ D_2 \\ D_3 \end{pmatrix} u, \tag{E.6}$$

where

$$C_1 = \left(-\dfrac{K_s}{M_s} \quad -\dfrac{B_s}{M_s} \quad 0 \quad \dfrac{B_s}{M_s} \right), \qquad D_1 = \dfrac{u_{max}}{M_s},$$
$$C_2 = (1 \ 0 \ 0 \ 0), \qquad\qquad\qquad D_2 = 0,$$
$$C_3 = (0 \ 0 \ 1 \ 0), \qquad\qquad\qquad D_3 = 0.$$

We consider the following values for the model parameters [102, 133]

$$M_s = 320 \ kg, \qquad\qquad K_s = 18 \ \dfrac{kN}{m},$$
$$B_s = 1 \ \dfrac{kN \cdot s}{m}, \qquad\qquad K_u = 200 \ \dfrac{kN}{m},$$
$$M_u = 40 \ kg, \qquad\qquad u_{max} = 1.5 \ kN,$$
$$SS = 0.08 \ m.$$

In a real control problem one has to take into account that, due to the actuators saturation, the active force is bounded by u_{max}, i.e., the normalized force has to satisfy

$$|u(t)| \leq 1, \qquad t \geq 0. \tag{E.7}$$

Eventually, the objective of the active suspension is to keep as small as possible the body acceleration $\ddot{x}_s(\cdot) = \dot{x}_2(\cdot)$ on a finite-time interval.

E.3 Inverted Pendulum

In this section we illustrate the inverted pendulum shown in Fig. E.4. Given the parameters reported in Table E.2, letting $x = (s \ \dot{s} \ \varphi \ \dot{\varphi})^T$ and $y = (s \ \varphi)^T$, the following

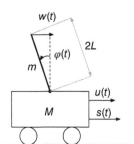

Figure E.4 Scheme of the inverted pendulum.

Table E.2 Parameters of the inverted pendulum.

s	cart position	
$\dot{s}$	cart velocity	
φ	angular position of the pendulum	
$\dot{\varphi}$	angular velocity related to φ	
u	control force applied to the cart	
w	disturbance force applied to the pendulum	
M	mass of the cart	0.5 *kg*
m	mass of the pendulum	0.2 *kg*
b	coefficient of friction of the cart	0.1 N/(m · sec)
L	length to the pendulum center of mass	0.3 m
I	mass moment of inertia of the pendulum	0.006 *kg* · m^2

linearized model can be derived

$$
\dot{x}(t) = \begin{pmatrix} 0 & 1 & 0 & 0 \\ 0 & -\dfrac{b(I+mL^2)}{I(M+m)+mML^2} & \dfrac{m^2L^2g}{I(M+m)+mML^2} & 0 \\ 0 & 0 & 0 & 1 \\ 0 & -\dfrac{bmL}{I(M+m)+mML^2} & \dfrac{mlg(M+m)}{I(M+m)+mML^2} & 0 \end{pmatrix} x(t)
$$

$$
+ \begin{pmatrix} 0 \\ \dfrac{I+mL^2}{I(M+m)+mML^2} \\ 0 \\ \dfrac{mL}{I(M+m)+mML^2} \end{pmatrix} u(t) + \begin{pmatrix} 0 \\ \dfrac{I-mL^2}{I(M+m)+mML^2} \\ 0 \\ \dfrac{mL-2L(M+m)}{I(M+m)+mML^2} \end{pmatrix} w(t) \qquad \text{(E.8a)}
$$

$$
z(t) = \begin{pmatrix} 1 & 0 & 0 & 0 \\ 0 & 0 & 1 & 0 \end{pmatrix} x(t). \qquad \text{(E.8b)}
$$

E.4 Yaw and Lateral Motions of a Two-Wheel Vehicle

A two-degrees-of-freedom model of a two-wheel vehicle with no roll, known as the *bicycle model*, is introduced in this section. This model describes both the lateral and yaw motions of the vehicle, neglecting the roll. It is worth to notice that this model does not describe the full behavior of a bicycle, but it is commonly used to describe the behavior of a four-wheel vehicle when small differences between the running condition of the left and the right side are ignored (see [134]). A comprehensive coverage of the bicycle dynamics can be found in [135].

We assume that the bicycle has only one steering wheel, the front one, and we denote by δ its steering angle. Moreover the longitudinal velocity v_x is assumed to be constant.

The two degrees of freedom that are considered are the lateral position y and the yaw ψ, being $\dot{y}$ the lateral velocity and $\omega := \dot{\psi}$ the yaw rate, both expressed in the body-fixed frame. Under the proper assumptions, the following linear model can be used to described the motion of the bicycle along the two considered degrees of freedom (see [136] for more details)

$$
\dot{x}(t) = \begin{pmatrix} -\dfrac{C_{af} + C_{ar}}{Mv_x} & -\dfrac{C_{af}l_f - C_{ar}l_r}{Mv_x} - v_x & 0 & 0 \\[2ex] -\dfrac{C_{af}l_f - C_{ar}l_r}{Iv_x} & -\dfrac{C_{af}l_f^2 + C_{ar}l_r^2}{Iv_x} & 0 & 0 \\[2ex] 0 & 1 & 0 & 0 \\[1ex] 1 & 0 & v_x & 0 \end{pmatrix} x(t) + \begin{pmatrix} \dfrac{C_{af}}{M} \\[2ex] \dfrac{C_{af}l_f}{I} \\[2ex] 0 \\[1ex] 0 \end{pmatrix} u(t),
$$

$$(\text{E.9})$$

with the following choice for the output

$$ y = \begin{pmatrix} \dot{y} & \omega & Y \end{pmatrix}^T, $$

the state

$$ x = \begin{pmatrix} \dot{y} & \omega & \psi & Y \end{pmatrix}^T, $$

and the control input

$$ u = \delta; $$

where

- Y is the vertical displacement of the bicycle center of gravity (CoG) in the ground-fixed axes;
- M is the mass of the vehicle;
- I is the yaw inertia;
- v_x is the longitudinal velocity, which is assumed to be constant;
- l_f and l_r are distances from the CoG to the front and rear axle, respectively (see Fig. E.5);
- C_{af} and C_{ar} are the effective cornering stiffnesses at the front and rear axle, respectively.

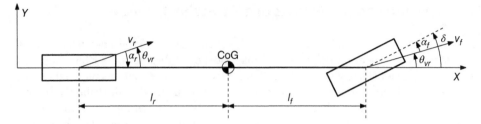

Figure E.5 Schematic representation of the bicycle, along with the various symbols adopted for its description, and the ground reference frame.

Table E.3 Two-wheel model parameters.

M	Mass of the vehicle	1200 kg
I	Yaw inertia	1500 kg·m²
C_{af}	Front cornering stiffness	125000 N/rad
C_{ar}	Rear cornering stiffness	80000 N/rad
l_f	Distance from the CoG and the front axle	0.92 m
l_r	Distance from the CoG and the rear axle	1.38 m
v_x	Longitudinal velocity	22 m/s
η	Side wind efficiency	0.15

Let us now include in (E.9) the exogenous input w in order to model the effect of the side wind on the lateral position Y. In particular, if w is the side wind velocity, the state equation (E.9) becomes

$$\dot{x}(t) = Ax(t) + Fw(t) + Bu(t), \qquad (E.10)$$

with

$$F = \begin{pmatrix} 0 & 0 & 0 & \eta \end{pmatrix}^T,$$

where η represents the *efficiency* of the side wind in moving the vehicle along the lateral direction.

The parameters for the bicycle model are reported in Table E.3 and have been mainly taken from [134].

References

1 Kamenkov, G. (1953) On stability of motion over a finite interval of time [in Russian]. *Journal of Applied Math. and Mechanics*, **17**, 529–540.

2 Lebedev, A. (1954) The problem of stability in a finite interval of time [in Russian]. *Journal of Applied Math. and Mechanics*, **18**, 75–94.

3 Lebedev, A. (1954) On stability of motion during a given interval of time [in Russian]. *Journal of Applied Math. and Mechanics*, **18**, 139–148.

4 Dorato, P. (1961) Short time stability in linear time-varying systems, in *Proc. IRE Int. Convention Record Pt. 4*, pp. 83–87.

5 Weiss, L. and Infante, E. (1967) Finite time stability under perturbing forces and on product spaces. *IEEE Transactions on Automatic Control*, **12**, 54–59.

6 Michel, A. and Porter, D. (1972) Practical stability and finite-time stability of discontinuous systems. *IEEE Transactions on Circuit Theory*, **CT-19**, 123–129.

7 Boyd, S., El Ghaoui, L., Feron, E., and Balakrishnan, V. (1994) *Linear Matrix Inequalities in System and Control Theory*, SIAM Press.

8 Amato, F., Ariola, M., and Dorato, P. (2001) Finite-time control of linear systems subject to parametric uncertainties and disturbances. *Automatica*, **37**, 1459–1463.

9 Amato, F., Ariola, M., Cosentino, C., Abdallah, C.T., and Dorato, P. (2003) Necessary and sufficient conditions for finite-time stability of linear systems, in *Proc. American Control Conference*, pp. 4452–4456.

10 Amato, F., Ariola, M., and Cosentino, C. (2006) Finite-time stabilization via dynamic output feedback. *Automatica*, **42**, 337–342.

11 Garcia, G., Tarbouriech, S., and Bernussou, J. (2009) Finite-Time Stabilization of Linear Time-Varying Continuous Systems. *IEEE Transactions on Automatic Control*, **54**, 364–369.

12 Amato, F., Ariola, M., and Cosentino, C. (2010) Finite-time stability of linear-time-varying systems: Analysis and controller design. *IEEE Transactions on Automatic Control*, **55**, 1003–1008.

13 Shen, Y. (2008) Finite-time control of linear parameter-varying systems with norm-bounded exogenous disturbance. *J. Contr. Theory Appl.*, **6** (2), 184–188.

14 Amato, F., Ariola, M., Carbone, M., and Cosentino, C. (2006) Finite-time control of linear systems: A survey, in *Current trends in nonlinear systems and control*, Birkhäuser, Boston.

15 Shaked, U. and Suplin, V. (2001) A new bounded real lemma representation for the continuous-time case. *IEEE Transactions on Automatic Control*, **46** (9), 1420–1426.

Finite-Time Stability: An Input-Output Approach, First Edition.
Francesco Amato, Gianmaria De Tommasi, and Alfredo Pironti.
© 2018 John Wiley & Sons Ltd. Published 2018 by John Wiley & Sons Ltd.

16 Davis, J., Gravagne, I., Marks, R., and Ramos, A. (2010) Algebraic and dynamic Lyapunov equations on time scales, in 42nd *Southeastern Symposium on System Theory (SSST)*, pp. 329–334.

17 Mastellone, S., Dorato, P., and Abdallah, C. (2004) Finite-time stability of discrete-time nonlinear systems: analysis and design., in *Proc. IEEE Conference on Decision and Control*, pp. 2572–2577.

18 Yang, Y., Li, J., and Chen, G. (2009) Finite-time stability and stabilization of nonlinear stochastic hybrid systems. *Journal of Mathematical Analysis and Applications*, **356**, 338–345.

19 Zhao, S., Sun, J., and Liu, L. (2008) Finite-time stability of linear time-varying singular systems with impulsive effects. *International Journal of Control*, **81** (11), 1824–1829.

20 Ambrosino, R., Calabrese, F., Cosentino, C., and De Tommasi, G. (2009) Sufficient conditions for finite-time stability of impulsive dynamical systems. *IEEE Transactions on Automatic Control*, **54** (4), 861–865.

21 Wang, Y., Shi, X., Wang, G., and Zuo, Z. (2012) Finite-time stability for continuous-time switched systems in the presence of impulse effects. *IET Control Theory and Applications*, **6**, 1741–1744.

22 Chen, G. and Yang, Y. (2012) Finite-time stability of switched positive linear systems. *International Journal of Robust and Nonlinear Control*, **24**, 179–190.

23 Amato, F., De Tommasi, G., and Pironti, A. (2013) Necessary and sufficient conditions for finite-time stability of impulsive dynamical linear systems. *Automatica*, **49** (8), 2546–2550.

24 Zuo, Z., Liu, Y., Wang, Y., and Li, H. (2012) Finite-time stochastic stability and stabilization of linear discrete-time Markovian jump systems with partly unknown transition probabilities. *IET Control Theory and Applications*, **6**, 1522–1526.

25 He, S. and Liu, F. (2010) Stochastic finite-time control for uncertain jump system with energy-storing electrical circuit simulation. *Int. J. Energy and Environment*, **1**, 883–96.

26 Luan, X.L., Liu, F., and Shi, P. (2010) Finite-time filtering for nonlinear stochastic systems with partially known transition jump rates. *IET Contr. Theory Applic.*, **4** (5), 735–745.

27 Yin, J., Khoo, S., Man, Z., and Yu, X. (2011) Finite-time stability and instability of stochastic nonlinear systems. *Automatica*, **42** (2), 337–342.

28 Yan, Z., Zhang, G., and Zhang, W. (2013) Finite-time stability and stabilization of linear Itô stochastic systems with state and control dependent noise. *Asian J. Control*, **15** (1), 270–281.

29 Yan, Z., Zhang, W., and Zhang, G. (2015) Finite-time stability and stabilization of Itô stochastic systems with Markovian switching: Mode dependent parameter approach. *IEEE Transactions on Automatic Control*, **60** (9), 2428–2433.

30 Amato, F., Ambrosino, R., Cosentino, C., and De Tommasi, G. (2010) Input-output finite-time stabilization of linear systems. *Automatica*, **46**, 1558–1562.

31 Amato, F., Carannante, G., De Tommasi, G., and Pironti, A. (2012) Input-output finite-time stability of linear systems: Necessary and sufficient conditions. *IEEE Transactions on Automatic Control*, **57** (12), 3051–3063.

32 Khalil, H.K. (1992) *Nonlinear Systems*, MacMillan Publishing Company.

33 Amato, F., Ariola, M., and Cosentino, C. (2005) Finite-Time Control of Linear Time-Varying Systems via Output Feedback, in *Proc. American Control Conference*, Portland, MD, pp. 4723–4727.

34 Amato, F., Ariola, M., Carbone, M., and Cosentino, C. (2006) Finite-time output feedback control of linear systems via differential linear matrix conditions, in *Proc. 45th IEEE Conference on Decision and Control*, San Diego, CA.

35 Amato, F., Ambrosino, R., Ariola, M., Cosentino, C., and De Tommasi, G. (2014) *Finite-Time Stability and Control*, Springer-Verlag.

36 Amato, F., Ariola, M., and Dorato, P. (1998) Robust finite-time stabilization of linear systems depending on parametric uncertainties, in *Proc. 37th IEEE Conference on Decision and Control*, Tampa, FL.

37 Amato, F., Ariola, M., Abdallah, C.T., and Dorato, P. (1999) Finite-time control for uncertain linear systems with disturbance inputs, in *Proc. American Control Conference*, pp. 1776–1780.

38 Amato, F., Ariola, M., Abdallah, C.T., and Dorato, P. (1999) Dynamic output feedback finite-time control of LTI systems subject to parametric uncertainties and disturbances, in *Proceedings of the 1999 European Control Conference*, Kalrsruhe (Germany), pp. 1776–80.

39 Amato, F., Ariola, M., Abdallah, C.T., and Cosentino, C. (2002) Application of finite-time stability concepts to the control of atm networks, in *Proceedings of the 40th Annual Allerton Conference on Communication Control and Computing*, Monticello, IL, pp. 1071–1079.

40 Amato, F. and Ariola, M. (2005) Finite-time control of discrete-time linear systems. *IEEE Transactions on Automatic Control*, **50**, 724–729.

41 Amato, F., Ambrosino, R., Ariola, M., and Calabrese, F. (2007) Finite-time stability of linear systems: an approach based on polyhedral lyapunov functions, in *Proc. IEEE Conf. on Decision and Control*, pp. 1100–1105.

42 Amato, F., Ambrosino, R., Ariola, M., and Calabrese, F. (2008) Finite-time stability analysis of linear discrete-time systems via polyhedral Lyapunov functions, in *Proc. American Control Conference*, pp. 1656–1660.

43 Amato, F., Ambrosino, R., Ariola, M., and Calabrese, F. (2010) Finite-Time Stability of Linear Systems: An Approach Based on Polyhedral Lyapunov Functions. *Control Theory Applications, IET*, **4**, 167–1774.

44 Ambrosino, R., Garone, E., Ariola, M., and Amato, F. (2012) Piecewise quadratic functions for finite-time stability analysis, in *Proc. IEEE Conf. on Decision and Control*, pp. 6535–6540.

45 Amato, F., De Tommasi, G., Mele, A., and Pironti, A. (2016) New conditions for annular finite-time stability of linear systems, in *Proc. IEEE Conf. on Decision and Control*.

46 Amato, F., Carbone, M., Ariola, M., and Cosentino, C. (2004) Finite-time stability of discrete-time systems, in *Proc. American Control Conference*, pp. 1440–1444.

47 Amato, F., Carbone, M., Ariola, M., and Cosentino, C. (2004) Control of Linear Discrete-Time Systems over a Finite-Time Interval, in *Proc. Conference on Decision and Control*, pp. 1284–1289.

48 Amato, F., Ariola, M., and Cosentino, C. (2005) Finite-time control of linear time-varying systems via output feedback, in *Proc. American Control Conference*, pp. 4722–4726.

49 Amato, F., Ariola, M., and Cosentino, C. (2010) Finite-time control of discrete-time linear systems: analysis and design conditions. *Automatica*, **46**, 919–924.

50 Amato, F., Darouach, M., and De Tommasi, G. (2016) Finite-time stabilizability and detectability of linear systems. part I: Necessary and sufficient conditions for the existence of output feedback finite-time stabilizing controllers, in *Proc. European Control Conference*, pp. 1412–1417.

51 Amato, F., Darouach, M., and De Tommasi, G. (2016) Finite-time stabilizability and detectability of linear systems. part II: Design of observer based output feedback finite-time stabilizing controllers, in *Proc. European Control Conference*, pp. 1406–1411.

52 Amato, F., Ariola, M., and Cosentino, C. (2011) Robust finite-time stabilization of uncertain linear systems. *International Journal of Control*, **84** (12), 2117–2127.

53 Amato, F., Cosentino, C., and Merola, A. (2010) Sufficient Conditions for Finite-Time Stability and Stabilization of Nonlinear Quadratic Systems. *IEEE Transactions on Automatic Control*, **55** (2), 430–434.

54 Amato, F., Ariola, M., and Cosentino, C. (2003) Finite-time control with pole placement, in *Proc. European Control Conference*.

55 Amato, F., Carannante, G., De Tommasi, G., and Pironti, A. (2012) Input-output finite-time stabilization with constrained control inputs, in *Proc. IEEE Conf. on Decision and Control*, pp. 5731–5736.

56 Amato, F., Carannante, G., De Tommasi, G., and Pironti, A. (2014) Input-output finite-time stabilization of linear systems with input constraints. *IET Control Theory and Applications*, **8** (14), 1429–1438.

57 Amato, F., Cosentino, C., De Tommasi, G., and Pironti, A. (2016) The mixed H_∞/FTS control problem: analysis and state feedback control. *Asian Journal of Control*, **18**, 828–841.

58 Zames, G. (1966) On the input-output stability of time varying nonlinear feedback systems—Part I: Conditions derived using concepts of loop gain, conicity, and positivity. *IEEE Transactions on Automatic Control*, **11**, 228–238.

59 Chen, T. and Francis, B. (1991) Input-output stability of sampled-data systems. *IEEE Transactions on Automatic Control*, **36**, 50–58.

60 Sontag, E. and Wang, Y. (1999) Notions of input to output stability. *Syst. Contr. Lett.*, **38**, 235–248.

61 Zhou, T. (2011) On nonsingularity verification of uncertain matrices over a quadratically constrained set. *IEEE Transactions on Automatic Control*, **56**, 2206–2212.

62 Amato, F., Ambrosino, R., Cosentino, C., De Tommasi, G., and Montefusco, F. (2009) Input-output finite-time stability of linear systems, in *Proc. Mediterranean Control Conference*, Thessaloniki, Greece, pp. 342–346.

63 Amato, F., Carannante, G., De Tommasi, G., and Pironti, A. (2011) Necessary and sufficient conditions for input-output finite-time stability of linear time-varying systems, in *Proc. 49th IEEE Conference on Decision and Control*, Orlando, FL.

64 Amato, F., Ambrosino, R., Ariola, M., and De Tommasi, G. (2011) Input to output finite-time stabilization of discrete-time linear systems, in *Proc. 2011 IFAC World Cong.*, MIlan, Italy, pp. 151–156.

65 Amato, F., Carannante, G., and De Tommasi, G. (2011) Input-output finite-time stabilization of a class of hybrid systems via static output feedback. *International Journal of Control*, **84** (6), 1055–1066.

66 Amato, F., Carannante, G., and De Tommasi, G. (2011) Finite-time stabilization of switching linear systems with uncertain resetting times, in *Proc. Mediterranean Control Conference*, pp. 1361–1366.

67 Amato, F., Ambrosino, R., Ariola, M., De Tommasi, G., and Pironti, A. (2017) On the finite-time boundedness of linear systems. *Automatica, submitted*.

68 Ichihara, H. (2009) Necessary and sufficient conditions for finite-time boundedness of linear continuous-time systems, in *Proceedings of the 2009 IEEE Conference on Decision and Control*, Shanghai, pp. 3214–19.

69 Du, H., Lin, X., and Li, S. (2010) Finite-time boundedness and stabilization of switched linear systems. *Kybernetika*, **46**, 870–889.

70 Wang, Y., Liu, Y., and Zuo, Z. (2014) Finite-time boundedness of switched delay systems: the reciprocally convex approach. *IET Control Theory Appl.*, **8** (15), 1575–1580.

71 Lin, X., Du, H., Li, S., and Zou, Y. (2013) Finite-time boundedness and finite-time l_2 gain analysis of discrete-time switched linear systems with average dwell time. *J. Franklin Institute*, **350** (4), 911–928.

72 Xu, J. and Sun, Y. (2013) Finite-time control of networked control systems with bounded packet dropout. *Journal of Material Science*, **629**, 840–44.

73 Bhat, S. and Bernstein, D. (2000) Finite-Time Stability of Continuous Autonomous Systems. *SIAM Journal on Control and Optimization*, **38** (3), 751–766.

74 Nersesov, S. and Haddad, W. (2007) Finite-time stabilization of nonlinear impulsive dynamical systems, in *Proc. European Control Conference 2007*, Kos, Greece, pp. 91–98.

75 Hong, Y., z. Jiang, and Feng, G. (2010) Finite-Time Input-to-State Stability and Applications to Finite-Time Control Design. *SIAM Journal on Control and Optimization*, **48** (7), 4395–4418.

76 Yosida, K. (1980) *Functional Analysis*, Springer-Verlag.

77 Naylor, A.W. and Sell, G.R. (1982) *Linear Operator Theory in Engineering and Science*, Springer-Verlag, New York.

78 Pettersson, S. (1999) *Analysis and Design of Hybrid Systems*, Ph.D. thesis, Chalmers University of Technology.

79 Nersesov, S. and Haddad, W. (2008) Finite-time stabilization of nonlinear impulsive dynamical systems. *Nonlinear Analysis: Hybrid Systems*, **2**, 832–845.

80 Medina, E. (2007) *Linear impulsive control systems: A geometric approach*, Ph.D. thesis, School of Electrical Engineering and Computer Science, Ohio University.

81 Medina, E. and Lawrence, D. (2009) State feedback stabilization of linear impulsive systems. *Automatica*, **45** (6), 1476–1480.

82 Liberzon, D. (2003) *Switching in Systems and Control*, Springer-Verlag, Berlin-Heidelberg.

83 Song, J., Niu, Y., and Jia, T. (2015) Input-output finite-time stabilization of nonlinear stochastic system with missing measurements. *Int. J. Systems Science*, **47** (12), 2985–2995.

84 S. Huang, H. R. Karimi, Z.X. (2013) Input-output finite-time stability of positive switched linear systems with state delays, in *Proc. 9th Asian Control Conference*, Istanbul, Turkey.

85 Xue, W. and Li, K. (2014) Input-output finite-time stability of time-delay systems and its application to active vibration control, in *Proc. IEEE Int. Conf. Automation Science and Engineering (CASE)*, Taipei, Taiwan.

86 Huang, S., Xiang, Z., and Karimi, H. (2014) Input-output finite-time stability of discrete-time impulsive switched linear systems with state delays. *Circuits Systems and Signal Processing*, **33** (1), 141–158.

87 Zhang, G., Trentelman, H., Wang, W., and Gao, J. (2017) Input–output finite-region stability and stabilization for discrete 2-d fornasini–marchesini models. *Systems & Control Letters*, **99**, 9–16.

88 Ma, H. and Jia, Y. (2011) Input-output finite-time stability and stabilization of stochastic Markovian jump systems, in *Proc. IEEE Conf. on Decision and Control*, pp. 8026–8031.

89 Lofberg, J. (2004) YALMIP: a toolbox for modeling and optimization in Matlab, in *Proc. IEEE Symposium on Computer-Aided Control System Design*, Taipei, Taiwan, pp. 284–289.

90 Gahinet, P., Nemirovski, A., Laub, A.J., and Chilali, M. (1995) *LMI Control Toolbox*, The Mathworks Inc.

91 Labit, Y., Peaucelle, D., and Henrion, D. (2002) SEDUMI INTERFACE 1.02: a tool for solving LMI problems with SEDUMI, in *Proc. of IEEE Int. Symp. Computer Aided Control System Design*, pp. 272–277.

92 Abdallah, C., Amato, F., and Ariola, M. (2001) Input-output stability, in *Encyclopedia of Electrical and Electronics Engineering*, Wiley.

93 Callier, F. and Desoer, C. (1991) *Linear System Theory*, Springer-Verlag.

94 Yosida, K. (1980) *Functional Analysis*, Springer-Verlag.

95 Wilson, D. (1989) Convolution and Hankel operator norms for linear systems. *IEEE Transactions on Automatic Control*, **34** (1), 94–97.

96 Amato, F. (2006) *Robust Control of Linear Systems Subject to Uncertain Time-Varying Parameters*, Springer-Verlag.

97 Anderson, B. and Moore, J. (1989) *Optimal Control: Linear Quadratic Methods*, Prentice-Hall.

98 Gahinet, P. (1996) Explicit controller formulas for LMI-based H_∞ synthesis. *Automatica*, **32**, 1007–1014.

99 Sana, S. and Rao, V. (2001) Robust control of input limited smart structural systems. *IEEE Transactions on Control Systems Technology*, **9** (1), 60–68.

100 Geromel, J., Peres, P., and Bernussou, J. (1991) On a convex parameter space method for linear control design of uncertain systems. *SIAM Journal on Control and Optimization*, **29**, 381–402.

101 Amato, F., Ariola, M., and Cosentino, C. (2003) Finite time control via output feedback: A general approach, in *Proc. IEEE Conference on Decision and Control*, Maui, HI, USA, pp. 350–355.

102 Chen, H. and Guo, K.H. (2005) Constrained $\mathcal{H}_\infty$ control of active suspensions: an LMI approach. *IEEE Trans. Contr. Sys. Technol.*, **13** (3), 412–421.

103 Zames, G. (1981) Feedback and optimal sensitivity: Model reference transformation, multiplicative seminorms and approximate inverses. *IEEE Trans. Auto. Contr.*, **26** (2), 301–320.

104 Francis, B., Helton, J., and Zames, G. (1984) $\mathcal{H}_\infty$-optimal feedback controllers for linear multivariable systems. *IEEE Trans. Auto. Contr.*, **29** (10), 888–900.

105 Francis, B. (1987) A Course in $\mathcal{H}_\infty$ Control Theory, Springer-Verlag.

106 Francis, B. and Doyle, J. (1987) Linear control theory with an $\mathcal{H}_\infty$ optimality criterion. *SIAM J. Control Optim.*, **25** (4), 815–844.

107 Glover, K. and Doyle, J. (1988) State space formulae for all stabilizing controllers that satisfy an $\mathcal{H}_\infty$-norm bound and relations to risk sensitivity. *Syst. Contr. Lett.*, **11** (3), 167–172.

108 Doyle, J., Glover, K., Khargonekar, P., and Francis, B. (1989) State-space solutions to standard $\mathcal{H}_2$ and $\mathcal{H}_\infty$ control problems. *IEEE Trans. Auto. Contr.*, **34** (8), 831–847.

109 Glover, K., D. J. N. Limebeer, J.D., Kasenally, E., and Safonov, M. (1991) A characterization of all solutions to the four-block general distance problem. *SIAM J. Control Optim.*, **29** (2), 831–847.

110 Uchida, K. and Fujita, M. (1989) On the central controller: Characterizations via differential games and LEQG control problems. *Systems & Control Letters*, **13** (1), 9–13.

111 Limebeer, D., Anderson, B., Khargonekar, P., and Green, M. (1989) A game theoretic approach to $\mathcal{H}_\infty$ control for time-varying systems, in *Proceedings of MTNS*, Amsterdam, The Netherlands, pp. 4723–4727.

112 Basar, T. and Bernhard, P. (1991) $\mathcal{H}_\infty$-*Optimal Control and Related Minimax Design Problems*, Birkhauser.

113 Tadmor, G. (1990) Worst-case design in the time domain: The maximum principle and the standard $\mathcal{H}_\infty$ control problem. *Math. Contr. Sign. Syst.*, **3** (4), 301–324.

114 Khargonegar, P., Nagpal, K., and Poola, K. (1991) $\mathcal{H}_\infty$ control with transients. *SIAM J. Control Optim.*, **29** (6), 1373–1393.

115 Tadmor, G. (1990) Input output norms in general linear systems. *Int. Journal of Control*, **51**, 911–921.

116 Jakubovič, V.A. (1977) The S-procedure in linear control theory. *Vestnik Leningrad Univ. Math.*, **4**, 73–93.

117 Xie, L., Fu, M., and Souza, C.D. (1992) $\mathcal{H}_\infty$ control and quadratic stabilization of systems with parameter uncertainty via output feedback. *IEEE Trans. Auto. Contr.*, **37** (8), 1253–1256.

118 Sznaier, M. and Bu, J. (1998) Mixed $l_1/\mathcal{H}_\infty$ control of MIMO systems via convex optimization. *IEEE Trans. on Auto. Contr.*, **43** (9), 1229–1241.

119 Ji, X., Sun, Y., Huang, Y., and Su, H. (2009) Mixed $L_1/\mathcal{H}_\infty$ control for uncertain linear singular systems. *J. Control Theory Appl.*, **7** (2), 134–138.

120 Zhang, J. and Makris, N. (2001) Rocking response of free-standing blocks under cycloidal pulses. *J. Eng. Mech.*, **127** (5), 473–483.

121 Pettersson, S. (1999) *Analysis and Design of Hybrid Systems*, Ph.D. thesis, Chalmers University of Technology.

122 Amato, F., De Tommasi, G., and Pironti, A. (2015) Necessary and sufficient conditions for input-output finite-time stability of impulsive dynamical systems, in *Proc. American Control Conference*, pp. 5998–6003.

123 Amato, F., De Tommasi, G., and Pironti, A. (2016) Input-output finite-time stabilization of impulsive linear systems: Necessary and sufficient conditions. *Nonlinear Analysis: Hybrid Systems*, **19**, 93–106.

124 Amato, F., Carannante, G., and De Tommasi, G. (2011) Input-output finite-time stability of switching systems with uncertainties on the resetting times, in *Proc. Mediterranean Control Conference*, pp. 1355–1360.

125 Lygeros, J., Tomlin, C., and Sastry, S. (1999) Controllers for reachability specifications for hybrid systems. *Automatica*, **35** (3), 349–370.

126 Liberzon, D. (2003) *Switching in Systems and Control*, Birkhäuser.

127 Hahn, H. (1967) *Stability of Motion*, Springer-Verlag, Berlin-Heidelberg.

128 Rouche, N., Habets, P., and Laloy, M. (1977) *Stability Theory by Lyapunov's Direct Method*, Springer-Verlag, New York.

129 Vidyasagar, M. (1993) *Nonlinear Systems Analysis*, Prentice-Hall International Editions.

130 Kalman, R.E. (1960) Control system analysis and design via the "Second Method" of Lyapunov - I continuous-time systems. *ASME J. Basic Engineering*, **82**, 371–394.

131 Pozo, F., Acho, L., Rodellar, J., and Rossell, J.M. (2009) A velocity-based seismic control for base-isolated building structures, in *Proc. American Control Conference*, St. Louis, MO, USA.

132 Kelly, J.M., Leitmann, G., and Soldatos, A. (1987) Robust control of base-isolated structures under earthquake excitation. *Journal of Optimization Theory and Applications*, **53** (1).

133 Gordon, T., Marsh, C., and Milsted, M. (1991) A comparison of adaptive LQG and nonlinear controllers for vehicle suspension system. *Vehicle Syst. Dyn.*, **20**, 321–340.

134 Sharp, R. and Valtetsiotis, V. (2001) Optimal preview car steering control. *Vehicle System Dynamics Supplement*, **35**, 101–117.

135 Jazar, R. (2008) *Vehicle Dynamics: Theory and Applications*, Springer.

136 Ariola, M., De Tommasi, G., and Amato, F. (2017) Vehicle collision avoidance via control over a finite-time horizon, in *2017 IEEE 14th International Conference on Networking, Sensing and Control (ICNSC)*, Calabria, Italy, pp. 222–227.

Index

Finite-Time Stability: An Input-Output Approach, First Edition.
Francesco Amato, Gianmaria De Tommasi, and Alfredo Pironti.
© 2018 John Wiley & Sons Ltd. Published 2018 by John Wiley & Sons Ltd.